Machine and Deep Learning Algorithms and Applications

Synthesis Lectures on Signal Processing

Editor
José Moura, *Carnegie Mellon University*

Synthesis Lectures in Signal Processing publishes 80- to 150-page books on topics of interest to signal processing engineers and researchers. The Lectures exploit in detail a focused topic. They can be at different levels of exposition—from a basic introductory tutorial to an advanced monograph—depending on the subject and the goals of the author. Over time, the Lectures will provide a comprehensive treatment of signal processing. Because of its format, the Lectures will also provide current coverage of signal processing, and existing Lectures will be updated by authors when justified.

Lectures in Signal Processing are open to all relevant areas in signal processing. They will cover theory and theoretical methods, algorithms, performance analysis, and applications. Some Lectures will provide a new look at a well established area or problem, while others will venture into a brand new topic in signal processing. By careful reviewing the manuscripts we will strive for quality both in the Lectures' contents and exposition.

DSP for MATLAB™ and LabVIEW™ IV: LMS Adaptive Filtering
Forester W. Isen
2009

DSP for MATLAB™ and LabVIEW™ III: Digital Filter Design
Forester W. Isen
2008

DSP for MATLAB™ and LabVIEW™ II: Discrete Frequency Transforms
Forester W. Isen
2008

DSP for MATLAB™ and LabVIEW™ I: Fundamentals of Discrete Signal Processing
Forester W. Isen
2008

The Theory of Linear Prediction
P. P. Vaidyanathan
2007

Nonlinear Source Separation
Luis B. Almeida
2006

Spectral Analysis of Signals: The Missing Data Case
Yanwei Wang, Jian Li, and Petre Stoica
2006

Machine and Deep Learning Algorithms and Applications

Uday Shankar Shanthamallu and Andreas Spanias

ISBN: 978-3-031-03748-1   paperback
ISBN: 978-3-031-03758-0   PDF
ISBN: 978-3-031-03768-9   hardcover

DOI 10.1007/978-3-031-03758-0

A Publication in the Springer series
SYNTHESIS LECTURES ON SIGNAL PROCESSING

Lecture #22
Series Editor: José Moura, *Carnegie Mellon University*
Series ISSN
Print 1932-1236   Electronic 1932-1694

Machine and Deep Learning Algorithms and Applications

Uday Shankar Shanthamallu
Arizona State University

Andreas Spanias
Arizona State University

SYNTHESIS LECTURES ON SIGNAL PROCESSING #22

ABSTRACT

This book introduces basic machine learning concepts and applications for a broad audience that includes students, faculty, and industry practitioners. We begin by describing how machine learning provides capabilities to computers and embedded systems to learn from data. A typical machine learning algorithm involves training, and generally the performance of a machine learning model improves with more training data. Deep learning is a sub-area of machine learning that involves extensive use of layers of artificial neural networks typically trained on massive amounts of data. Machine and deep learning methods are often used in contemporary data science tasks to address the growing data sets and detect, cluster, and classify data patterns. Although machine learning commercial interest has grown relatively recently, the roots of machine learning go back to decades ago. We note that nearly all organizations, including industry, government, defense, and health, are using machine learning to address a variety of needs and applications.

The machine learning paradigms presented can be broadly divided into the following three categories: supervised learning, unsupervised learning, and semi-supervised learning. Supervised learning algorithms focus on learning a mapping function, and they are trained with supervision on labeled data. Supervised learning is further sub-divided into classification and regression algorithms. Unsupervised learning typically does not have access to ground truth, and often the goal is to learn or uncover the hidden pattern in the data. Through semi-supervised learning, one can effectively utilize a large volume of unlabeled data and a limited amount of labeled data to improve machine learning model performances. Deep learning and neural networks are also covered in this book. Deep neural networks have attracted a lot of interest during the last ten years due to the availability of graphics processing units (GPU) computational power, big data, and new software platforms. They have strong capabilities in terms of learning complex mapping functions for different types of data. We organize the book as follows. The book starts by introducing concepts in supervised, unsupervised, and semi-supervised learning. Several algorithms and their inner workings are presented within these three categories. We then continue with a brief introduction to artificial neural network algorithms and their properties. In addition, we cover an array of applications and provide extensive bibliography. The book ends with a summary of the key machine learning concepts.

KEYWORDS

artificial intelligence, machine learning, deep learning, neural networks, Internet of things, supervised learning, unsupervised learning, signal processing, big data

Contents

Preface

This book was motivated in part by a series of independent study sessions and machine learning seminars of the SenSIP center at Arizona State University. The authors worked with a team of Ph.D. students and several industry members of the SenSIP center on a variety of projects in machine learning. A milestone in this effort was a tutorial prepared by the authors and presented in the SensMACH industry-university meeting in 2016. An expanded version of this tutorial was presented at the IEEE IISA 2017 international conference in Cyprus and a mini survey paper was published on IEEE Xplore. The authors also worked on a complementary series of machine learning education software applications which were presented at various conferences including IEEE ICASSP. Some of this content was also developed to support NSF workforce development programs such as the REU, RET, and IRES. Knowledge gained from research projects with our collaborators in Lawrence Livermore Laboratory also motivated certain algorithms described in this book.

The primary objective of this book is to introduce basic machine learning concepts and applications for use in academia, government labs, and industry. It is intended for a broad spectrum of readers, including undergraduate and graduate students, faculty, and industry/government practitioners. Our objective is to cover an array of concepts and applications in machine learning and deep learning. The book is self-contained and discusses several machine learning algorithms, including artificial neural networks and their training, and regularization. Toward the end, the reader will be exposed to emerging applications of deep learning for various data domains including vision, text, speech, relational data, energy, and natural language. An extensive literature review is provided in all the chapters with recommendations for further reading.

Uday Shankar Shanthamallu and Andreas Spanias
September 2021

Acknowledgments

The study was supported in part by various NSF awards. Logistical support was provided by the ASU SenSIP center which is an NSF I/UCRC site. We like to thank Mike Stanley of NXP and Jayaraman Thiagarajan of Lawrence Livermore National Labs, and our SenSIP center industry partners for encouraging us to write this book. We owe an outstanding debt of gratitude to Mike Stanley for extensive discussions and feedback on machine learning and co-teaching the initial short course with the authors. We also like to thank Ph.D. students in our lab for frequent discussions and presentations. The authors would like to express their appreciation to several research sponsors including NXP, ON Semi, Qualcomm, Raytheon, and NSF. We cite specifically NSF awards 1525716, 1550040, 1659871, 1854273, and 1953745 for their support of the authors at different stages of this project.

Uday Shankar Shanthamallu and Andreas Spanias
September 2021

CHAPTER 1

Introduction to Machine Learning

Machine Learning (ML) [1–4] is a field of science that deals with learning patterns from data features and statistics, without explicit rule-based programming. ML provides an important capability for computers to learn from data examples and experience. A few simple machine learning examples include image classification [5, 6], spam email filtering [7], and stock price prediction [8]. For image classification such as dog vs. cat, an ML model is trained on thousands of images of dogs and cats until it can learn to distinguish the two. Similarly, for spam email filtering, an ML model can be trained with a lot of benign and spam emails to filter future spam messages.

On the other hand, deep learning [9, 10] is a sub-area of machine learning that involves extensive use of layers of artificial neural networks (ANNs) [11] that are trained on massive amounts of data. Here data broadly refers to all the digital content available in the form of images, text, speech, biometrics, DNA, health measurements, surveillance, user preferences, mouse clicks, etc. In fact, an average person may generate thousands of bits of digital data in every activity: be it reading an article, sending an email, tagging an image on Facebook, tweeting, exercising, or even watching television.

Many segments of industries employ machine learning and deep learning algorithms to determine patterns in data that will provide enhanced user experiences. For example, Google uses ML to rank web pages [12]. Amazon employs ML services in its back-end to provide better recommendations [13] according to the customer's purchase and browsing history. Netflix provides movie recommendations [14] according to watching habits. ML and Internet-of-Things (IoT) applications are becoming mainstream with several products, including the Amazon Echo, Google Home™, and Apple HomePod™ entering households. These products embed advanced signal processing, cloud computing, sensor arrays, and machine learning, enabling voice recognition with elevated accuracy.

Artificial Intelligence (AI), Machine learning, and Deep Learning (DL) are currently "hot" areas with a myriad applications and the DL area received large R&D investments during the last decade. Although commonalities and intersections exist in the aforementioned fields, each field is different in a broader context. The field of machine learning algorithms and applications is vast and is continuously expanding. In a nutshell, machine learning deals with identifying patterns in data and making predictions.

1.1   BRIEF HISTORY

Although machine learning and deep learning gained popularity in the last decade, ML concepts emerged in the early 20th century. Interestingly, this was also the inception of the most basic unit in the artificial neural networks: the perceptron. On an interesting note, statisticians might disagree because early machine learning algorithms focused on studying the underlying statistical patterns in the data and employed previously existing statistical concepts to learn from data. For example, the least squares regression and Bayes' theorem [15] were already known in the 19th century. These are now widely known as linear regression and naive Bayes rule, respectively.

Earlier machine learning algorithms consisted of purely mathematical models built to explain data or learn from data. Once computers were invented, researchers began discussing whether the machines would eventually acquire human-like intelligence. To this end, Alan Turing proposed the Turing test [16] for assessing a machine's ability to exhibit human-like intelligence. In 1957, Frank Rosenblatt invented the perceptron [17], a basic unit in the neural network, inspired by the logic gates that were popular at the same time. However, at that time, the limitations of neural networks [18], such as lack of training algorithms and lack of computational hardware, hindered their research.

In the last quarter of the 20th century, researchers shifted their focus on machine learning ideas from a knowledge-driven approach to a data-driven approach. Researchers started creating computer programs that learned from data directly by mapping input features to output decision. For example, in digital telephony, codebook vectors were proposed using vector quantization [19] that enables data compression. Many machine learning algorithms that relied on statistical modeling and optimization were invented. Algorithms such as support vector machines (SVM) and random forests were introduced. These algorithms were sufficient to deal with a relatively small amounts of data. Nevertheless, with the advent of the Internet, multi-sensor systems and broadband connectivity, digital data has grown tremendously.

From 2010 onward, computing capabilities needed to train deep learning models to improve their discriminative power became available and thus DL began gaining increasing interest in several application domains. During the decade of 2010–2020, we witness several innovations in deep learning algorithms, architectures, and implementations. During the same time, several algorithms that accelerated deep neural network training, such as batch normalization, dropout, and layer normalization, were proposed. New generative modeling techniques such as Variational Auto-Encoders (VAE) [20] and Generative Adversarial Networks (GAN) [21], which advanced the field of unsupervised learning are also based on innovations on deep neural net architectures. In addition, several processor architectures [22] were designed specifically for deep learning. During this last decade we also witnessed deep learning innovations in many fields, including natural language processing, speech processing, autonomous vehicles, medical diagnostics, drug discovery, and many security applications. A summary of the milestones in machine learning and deep learning in terms of a timeline is shown in Figure 1.1.

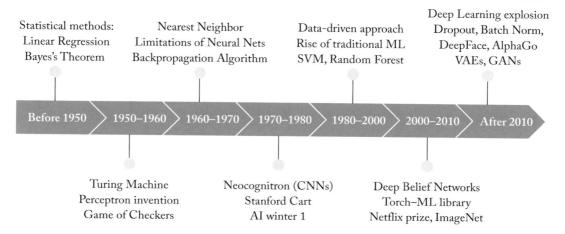

Figure 1.1: A timeline of machine learning and deep learning milestones.

1.2   LEARNING PARADIGMS

In general, learning is an active process that builds on prior knowledge, and it provides the ability to reason. With machine learning, computers can actively learn from data and understand the underlying knowledge embedded in the very data itself. Traditionally, computer programs and programming languages have been hugely successful in abstracting real-world interactions. Programmers were able to break down complex tasks into a set of modules that implemented a set of operations following rules.

Why is machine learning needed?
Computers excel in performing massive mathematical calculations and numerical operations that can be broken down into simple steps. Nevertheless, it is still hard to solve indefinite integrals or nonlinear partial differential equations that do not have numerical methods established. Although computers perform complex mathematical operations performing tasks such as object identification and recognition, and face identification is still a challenge. These difficulties in vision and image recognition are due to the limitations in algorithms and programming constructs. For example, it is hard to write a program that solves problems such as recognizing three-dimensional objects under various illumination patterns. In addition, it is extremely difficult to identify a set of rules that would work for all possible scenarios, such as the angle of viewing, varying illumination, scale, and occlusion. With machine learning, computers no longer use a fixed set of predefined hard-coded rules. Instead, a ML model is trained using large amounts of data and learns how to identify patterns and make predictions from the data directly.

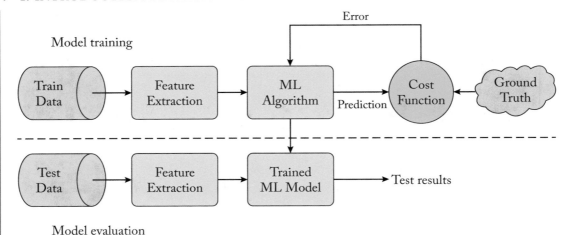

Figure 1.2: Training and evaluating Machine Learning algorithms using ground truth data.

How does machine learning work?

The goal of machine learning is to learn from examples (data) and experiences. Tom Mitchell gave a "well-posed" formal definition for machine learning [23]: "A computer program is said to learn from experience E with respect to some task T and some performance measure P, if its performance on T, as measured by P, improves with experience E."

The way the machine can learn depends on the learning paradigm itself. Machine learning algorithms can be broadly classified into *Supervised* and *Unsupervised*. Supervised learning, as the name suggests, has some form of supervision available with the data. This supervision is often in the form of using *Ground Truth* or *True Labels* contained in the data. The ML model is "trained" using the labeled input dataset (train data) to make predictions. The term "train" refers to the process where the ML model's parameters are updated to make better predictions.

During the training process, the ML model makes appropriate predictions on the input data and improves its prediction estimates using the ground truth and reiterating until the model reaches the desired accuracy level. Once the model has achieved the desired accuracy, it can then be used to perform inference on unseen data. We optimize the model parameters according to a cost function or an objective function in almost all the machine learning algorithms. The cost function is typically a measure of the error between the ground truth and the model predictions. By minimizing the cost function, we train our ML model to produce estimates close to the ground truth. Minimization of the cost function is usually achieved using gradient descent techniques (see Chapter 5). The training process is also known as *empirical risk minimization*. Variants of gradient descent techniques such as stochastic gradient descent for minibatch [24], and momentum-based gradient descent [25], Nesterov accelerated gradient descent [26], have been used in different training paradigms. Supervised learning can be again classified into sub-categories (see Chapter 2).

On the other hand, there is no explicit supervision or ground truth associated with the data in unsupervised learning. The objective is to draw inferences from the input data and then model the hidden or the underlying structure and the data distribution to learn more about the data. Chapter 3 describes unsupervised learning paradigms and different algorithms. Note that this book does not cover reinforcement learning, another paradigm in machine learning. Readers are recommended to go through [27–29] for topics in reinforcement learning.

1.3   THE EMERGENCE OF DEEP LEARNING

Although machine learning algorithms have been extensively used since the 1990s, deep learning became popular in the early 2010s. Before deep learning, engineers and domain experts spent several years developing feature descriptors, for example, Mel-frequency cepstral coefficients (MFCC) features [30] for speech and audio applications, scale-invariant feature transforms (SIFT) [31], and histogram of oriented gradients (HOG) [32] for image and vision applications. Note that feature descriptors "distill" patterns out of what otherwise might be indecipherable raw data. These feature descriptors for data were used to train known ML algorithms such as support vector machines or random forest classifiers to make predictions. Hand-designed feature descriptors are not needed for deep learning algorithms. Instead, raw data (image, audio, for example) are fed to the DL models which are tasked to learn the discriminative features by themselves. The rise in popularity of deep learning can be mainly attributed to the following.

1. **Big Data:** With the advent of the Internet, multi-sensor systems, broadband connectivity, and social media platforms, the creation and utilization of digital content became a new norm, leading to an abundance of data. Furthermore, a massive amount of data is required to train deep learning models to help interpret and provide analytics.

2. **Computing Capabilities**: Parallel computing devices in the form of Graphics Processing Units (GPUs) became compact and affordable. DL algorithms train on massive amounts of data and hence require very high computational resources to achieve this. Training is often performed in the cloud, and providers such as Microsoft, Google, Amazon, and others provide scalable hardware frameworks to enable machine learning model development and deployment.

3. **Software Platforms**: Many deep learning tools, for example, TensorFlow and PyTorch, became available for various software platforms and frameworks. These frameworks helped researchers accelerate deep learning model development, prototyping, and deployment.

1.4   RELATION BETWEEN AI, ML, AND DL

As mentioned earlier, AI, ML, and DL are often used interchangeably. However, there is a specific hierarchy among the three. AI, composed of two words: *Artificial* and *Intelligence*, is any unnatural system or a framework, a machine, or an agent that can possess and exhibit human-like intelligence. AI can be subdivided into Artificial Narrow Intelligence and Artificial General Intelligence [33]. Most of the existing progress seen is in the former field. Artificial Narrow Intelligence is the sophisticated set of specialized frameworks such as those used in self-driving cars, digital voice assistants with automatic speech recognition, intelligent surveillance, etc.

On the other hand, Artificial General Intelligence refers to the systems which possess human-like intelligence, toward which limited progress is seen. AI is a vast field, and machine learning is only a subset of AI where learning algorithms parse and learn from data. ML algorithms and models allow computers to learn from examples and experience. ML models carry out a specific task in the AI framework. DL is a subset of machine learning which heavily relies on layers of artificial neural networks and massive amounts of data for training. While DL models learn features as part of the training process, classical ML models utilize hand-crafted features for training and predictions. The hierarchical relationship among AI, ML, and DL, along with respective examples, is shown in Figure 1.3.

1.5   ORGANIZATION OF THE BOOK

The book is written so the reader is first introduced to general concepts in machine learning. It then highlights commonly used practical ML algorithms in each learning paradigm. Important notation is defined throughout the text and terms are defined and often highlighted using *italics*. At the end of each chapter, a summary section is provided to highlight the important concepts presented in that chapter.

Chapter 2 introduces concepts in supervised machine learning. It also discusses the sub-categories in supervised learning. It focuses on commonly used regression algorithms, including linear and non-linear regression. In addition, it covers ML classification problems while providing an understanding of common classification algorithms such as support vector machines, K-nearest neighbor, decision trees, etc.

Chapter 3 discusses the importance of unsupervised learning in the absence of true labels to the data. It introduces different styles of models often employed in unsupervised learning. The commonly used clustering algorithms such as the K-means algorithm, the spectral clustering, and the Gaussian Mixture Models are discussed. The last section of the chapter deals with dimensionality reduction algorithms in machine learning.

Chapter 4 introduces semi-supervised learning. It explains how a large amount of unlabeled data can be used with a few labeled ones. Most of the chapter focuses on graph-based machine learning while showcasing different approaches in the literature. Positive unlabeled

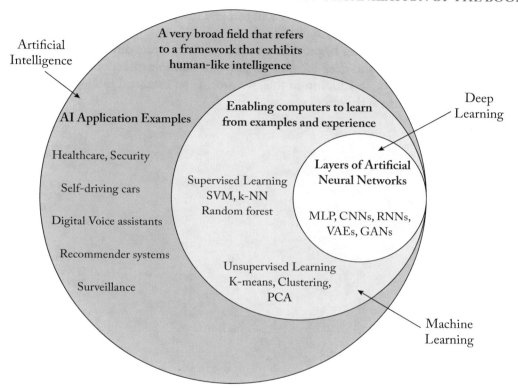

Figure 1.3: Relationship between Artificial Intelligence, Machine Learning, and Deep Learning.

learning, a sub-paradigm of semi-supervised learning, is also introduced at the end of the chapter.

In Chapter 5, a brief introduction to the field of artificial neural networks (ANNs) is provided with a focus on deep learning, neural network training, and different architectures. It starts with a perceptron discussion, a fundamental unit in neural networks. We show how the stacking of neurons in a layer-wise fashion leads to powerful ML models. We discuss how a neural network is trained through a backpropagation algorithm in the later sections. We also introduce and explain commonly used activation functions in a neural network. Different techniques employed to avoid over-fitting ANNs are discussed. Next, we discuss numerous ANN architectures used in practice for different data types. Finally, the last section of the chapter deals with unsupervised representation learning using neural nets.

Chapter 6 is dedicated to current ML and DL applications. We introduce several applications of machine learning and deep learning in different data domains, including sensor and time-series data, energy, image and vision, text and natural language processing, relational data, and Internet-of-Things (IoT) applications.

Chapter 7 concludes by providing a summary and future directions to expand knowledge on machine learning and deep learning. It provides multiple directions to follow depending on the reader's interests. In addition to this, we provide an extended bibliography throughout the book citing relevant authors and works.

CHAPTER 2

Supervised Learning

The supervised learning paradigm [34] is perhaps the most popular method in the machine learning community. In supervised learning, one has access to the ground truth for samples contained in the training, validation, and test data sets. Ground truth represents "true" or "correct" labels for the input dataset. Expert help may be needed to obtain the correct labels for the data (medical image labeling, for example). The ML model is "trained" using a labeled input dataset termed *training data*. Once the model achieves the desired performance on training data, the trained model is then used to perform inference on unseen data. The data that has not been used for training and thus unseen by the model is termed *test data*.

Let a set $\mathcal{X}$ represent a population of training data, and a set $\mathcal{Y}$ defines its corresponding ground truth. A supervised ML model typically aims to learn a mapping function f from the input to the true labels.

$$f : \mathcal{X} \mapsto \mathcal{Y}$$

Since the ML model is trained under the supervision of the ground truth, the learning process is termed *supervised learning*. Suppose we have N number of training examples and each one of them being labeled, the dataset can be represented in a pair $\{\mathbf{x}^{(i)}, y^{(i)}\}_{i=1}^{N}$ here $\mathbf{x}^{(i)}$ represents the ith training example. The input data can be a d dimensional vector $(\mathbf{x} \in \mathbb{R}^d)$ where each dimension corresponds to a feature or an independent variable. As an example, consider the problem of dog vs. cat image classification. For simplicity, we can consider grayscale images of size 28×28. In this case, images are represented by $\mathbf{x}$ and d is $784(28 \times 28)$ representing the pixel intensities. The ground truth y represents the correct label for the images. Mathematically, this can be represented by a number: 0 for dogs and 1 for cats.

Now, the goal is to train a ML model $f_{\boldsymbol{\theta}}$, where $\boldsymbol{\theta}$ represents the learnable parameters of the model. That is, we want to make predictions ($\hat{y}$) on the input images

$$\hat{y} = f_{\boldsymbol{\theta}}(\mathbf{x})$$

We then minimize the error in the predictions $\hat{y}$ and ground truth y. Minimization is usually based on a cost function or loss function $L(y, \hat{y})$, which is a measure of the error in the model's predictions. We iteratively compute the model's loss L and update the model parameter $\boldsymbol{\theta}$ to reduce this error. This process is summarized in Algorithm 2.1.

Algorithm 2.1 Supervised Learning Algorithm

Input: Training Dataset $\{\mathbf{x}^{(i)}, y^{(i)}\}_{i=1}^{N}$ and Model $f_{\boldsymbol{\theta}}$
Output: Trained Model $f_{\boldsymbol{\theta}}^{*}$
 1: **while** Model achieves required Performance **do**
 2:    Compute loss $L(y^{(i)}, \hat{y}^{(i)})$; $i \leftarrow 1$ to N
 3:    Update $\boldsymbol{\theta}$ to minimize L
 4: **end while**

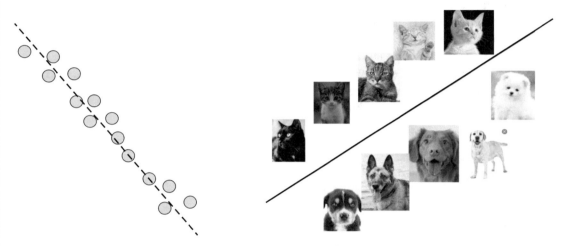

Figure 2.1: Difference between regression and classification. Regression deals with curve fitting (left image) whereas classification deals with learning a decision boundary (right image).

2.1   REGRESSION VS. CLASSIFICATION

As stated before, in supervised learning, a model is trained iteratively to learn the mapping function f from the training samples $\mathbf{X}$ (matrix form for $\mathbf{x}$) to the output variable $\mathbf{y}$; that is, $\mathbf{y} = f(\mathbf{X})$. The goal is to approximate the mapping function f as accurately as possible. Whenever there is a new input data point $\mathbf{x}$, the mapping function f can predict the output variable $\hat{y}$ for the dataset. Supervised learning is further divided into regression and classification algorithms [35–37]. Both share the same concept of utilizing ground truth to train an algorithm and make better predictions. Figure 2.1 shows a representative example for classification and regression methods.

In classification-based predictive modeling [38], the goal is to obtain a discrete outcome for the given training examples $\mathbf{x}$ using the mapping function f. For this reason, the output variables are often termed *labels* or *categories*. Thus, the mapping function predicts the class or category for a given observation. The cross-entropy function [334] is often used as a loss function to train the model to make better predictions. For example, an email can be classified as belonging to one of two classes: "spam" and "not spam" (*binary classification*). The number of outcomes is not necessarily limited to two categories. For example, an image of a bird can be classified into one of the hundred known categories (*multi-class classification*). In this case, y can take values from 0–99, one for each bird species.

In regression-based predictive modeling [37], one is interested in learning a mapping of the input variable to a continuous output variable. The dimensions of the input and output variables need not be the same. In this case, y is a continuous real-valued output, an integer, or a floating-point value. For example, consider training an ML model to predict the heights of individuals given their demographics. Simple loss functions such as mean-squared error (MSE) [39] or mean absolute error (MAE) [40] can be employed to train the ML model.

2.2   COMMON REGRESSION ALGORITHMS

In this section, we explain some of the commonly used supervised ML algorithms.

2.2.1   LINEAR REGRESSION

Linear regression [41], as the name suggests, is a linear ML model (linear function) which is purely a statistical technique of learning a mapping function between independent variables and dependent variables. The simplest example is the univariate linear regression [42] which consists of one independent variable (input) and one dependent variable (output) as shown below (note that x is a scalar):

$$\hat{y} = f(x) \tag{2.1}$$
$$\hat{y} = \theta_0 + \theta_1 x \tag{2.2}$$

Here, θ_0 and θ_1 are the parameters of the model f which represent the *intercept* and *slope* of the algebraic line, respectively. The model parameters are learnable, and they are iteratively updated to produce the best fit for the given data points. An intuitive figure showing the linear dependency between the input and the output variable is shown in Figure 2.2. A more sophisticated linear regression with multiple independent variables ($\mathbf{x} \in \mathbb{R}^d$) is shown below. Note that output y is still a linear function of the input variables ($x_1, x_2, \ldots, x_d$).

$$\hat{y} = f(\mathbf{x}) \tag{2.3}$$
$$\hat{y} = \theta_0 + \theta_1 x_1 + \theta_2 x_2 + \cdots + \theta_d x_d \tag{2.4}$$

The above equations represent a multi-variate linear regression [43] indicative of the multiple independent variables. The output y in both the cases is shown as a scalar for ease of understanding. However, in practice, y can also be multi-dimensional.

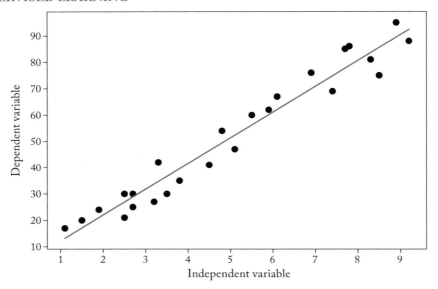

Figure 2.2: A linear regression between independent input variable and the dependent output variable.

2.2.2   NONLINEAR REGRESSION

Similar to linear regression, nonlinear regression [44] or polynomial regression [45] aims at predicting a continuous output variable. However, the nature of the relationship between the model's prediction and the input variables is no longer limited to be a linear function. The output variable is expressed as a sum of polynomial degrees of the input variable. A cubic-regression is shown in Figure 2.3. Alternately, y can also be expressed as a sum of other nonlinear functions such as exponential or logarithmic.

$$\hat{y} = \theta_0 + \theta_1 x + \theta_2 x^2 + \theta_3 x^3 \tag{2.5}$$
$$\hat{y} = \theta_0 + \theta_1 x + \theta_2 \log x \tag{2.6}$$

The model parameters θ weigh the input variables and their polynomial powers differently and effectively implement the resulting mapping function f.

For both linear and nonlinear regression algorithms, the MSE loss function is often used to tune and update the model parameters to make better predictions. The loss function L is given by Equation (2.7). The current model parameters θ is updated using the computed empirical loss. After calculating the loss, we compute the influence of θ on the final loss L, commonly referred

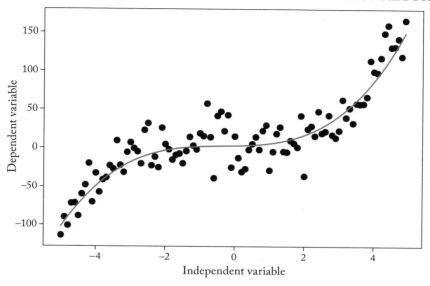

Figure 2.3: A cubic regression between independent input variable and the dependent output variable.

to as *gradient*, represented as $\frac{\partial L}{\partial \theta}$.

$$L(y, \hat{y}) = \frac{1}{N} \sum_{i=1}^{N} (\hat{y} - y)^2 \tag{2.7}$$

$$\theta^+ \leftarrow \theta \text{ and } \frac{\partial L}{\partial \theta}$$

2.3   COMMON CLASSIFICATION ALGORITHMS

2.3.1   LOGISTIC REGRESSION

Logistic regression [46, 47] is a classification algorithm used for predicting discrete outcomes from the input variables. Mathematically, logistic regression models a binomial outcome, which outputs a probability value mapped to two or more discrete classes. Binary logistic regression performs a binary classification of positive class from negative class. Logistic regression uses a sigmoid curve shown in Figure 2.4, which maps any real-value to a probability value between 0 and 1, and classification can be done with the help of a threshold.

$$S(z) = \frac{e^z}{1 + e^z} = \frac{1}{1 + e^{-z}} \tag{2.8}$$

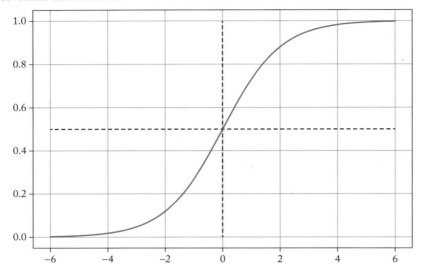

Figure 2.4: Sigmoid function showing the "S" shaped curve in the regular data regime.

$$\hat{y} = \begin{cases} 1 & \text{if } S(f(\mathbf{x})) > 0.5 \\ 0 & \text{if } S(f(\mathbf{x})) \leq 0.5 \end{cases} \tag{2.9}$$

Here, $f(\mathbf{x})$ is similar to Equation (2.4). Note that the class prediction $\hat{y}$ is made using the probability output from the sigmoid function and using the threshold as 0.5.

Binary cross-entropy (BCE) loss is used to measure the discrepancy between model prediction $\hat{y}$ and ground truth y, BCE is given as follows:

$$L(y, \hat{y}) = -\Big(y \log(\hat{y}) + (1 - y) \log(1 - \hat{y})\Big) \tag{2.10}$$

In simple terms, cross-entropy measures the difference between two probability distributions. In this case, the two probability distributions are obtained from ground truth y and predictions $\hat{y}$.

For a multi-class classification, a softmax function is used in place of logistic sigmoid function. Consider a K-class classification problem where we want the model to predict one of the K classes. The softmax function given by Equation (2.11) outputs a normalized probability score for all the classes. The outcome with the maximum score is chosen as the predicted class. To achieve this, the ML model employs K different θ parameter sets, represented in matrix form, we have $\Theta = [\theta_1, \theta_2, \cdots \theta_K]^T$ and $\hat{\mathbf{y}} = f_\Theta(\mathbf{x})$. Note that $\hat{\mathbf{y}}$ is a K-dimensional vector, representing probability for each of the K classes. The probability that $\mathbf{x}$ is assigned to kth class

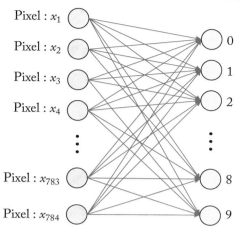

Figure 2.5: A multi-class classification using logistic regression with softmax function applied to the outputs.

is given by

$$p(\hat{y} = C_k \mid \mathbf{x}, \mathbf{\Theta}) = \frac{e^{h_k}}{\sum_{j=1}^{K} e^{h_j}}; \quad h_j = f_{\boldsymbol{\theta}_j}(\mathbf{x}) \;\; j = 1, 2, \ldots, K \tag{2.11}$$

To train and update the multi-class ML model parameters $\mathbf{\Theta}$, a negative log-likelihood function similar to Equation (2.10) is used.

An application of the logistic regression algorithm to handwritten digits from the MNIST dataset is discussed below. Each image in the dataset has 28×28 pixel resolution and is vectorized to get a 784 dimension variable. After training the logistic model in the Figure 2.5, the parameters (also referred to as weights) for each class are visualized and shown in Figure 2.6. It is evident that the model is learning a template for each of the handwritten digits from the entire training data and hence may not provide the best performance.

2.3.2   SUPPORT VECTOR MACHINES

Support Vector Machines (SVMs) [48] are one of most popular non-probabilistic supervised learning algorithms. Although SVM is mainly used for classification purposes it can also be used in regression [49] and clustering settings [50]. SVM when applied for binary-classification explicitly learns a *decision boundary* that separates data based on their ground truth. Mathematically, the decision boundary is known as a *hyperplane*. An example of SVM for binary-classification is shown in Figure 2.7. The data points belonging to the positive class ($\mathbf{x}_+$) and the negative class ($\mathbf{x}_-$) can be separated by many decision boundaries (indicated with dotted gray lines). A decision boundary can be mathematically defined with a hypothesis function

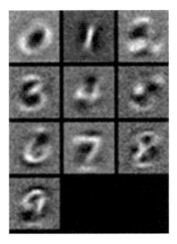

Figure 2.6: Parameters of the multi-class logistic model visualized after training.

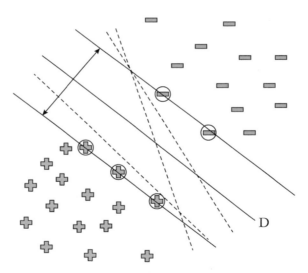

Figure 2.7: The SVM algorithm for binary classification. SVM enables a maximum margin between two classes. Support vectors which are on the immediate vicinity of the decision boundary are shown with circles.

$h(\mathbf{x}) = \boldsymbol{\theta} \cdot \mathbf{x} + b$ and model prediction as follows:

$$\hat{y} = \begin{cases} 1 & \text{if } h(\mathbf{x}) > 0 \\ 0 & \text{if } h(\mathbf{x}) < 0 \end{cases} \qquad (2.12)$$

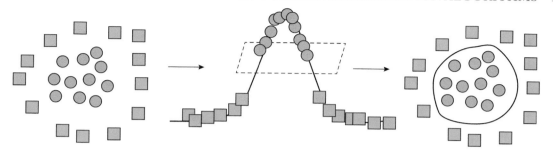

Figure 2.8: Shown here is the kernel trick achieved using a Gaussian kernel. The input data is transformed into a higher-dimensional space. A linear decision boundary (a 2D plane) is learned and transformed back to the original feature space to learn a nonlinear decision boundary.

The model parameters $\boldsymbol{\theta}$ and bias b are mutable and updated using the data points. Different values of $\boldsymbol{\theta}$ and bias b lead to different decision boundaries. Among the many decision boundaries, which one is the best or more optimal? SVM provides the best solution to the previous question. SVM constructs a decision boundary (D in the Figure 2.7) that provides the maximum margin, i.e., the largest distance between the hyperplane and training data points of either class. Ideally, $\boldsymbol{\theta} \cdot \mathbf{x}_+ + b > 0$ and $\boldsymbol{\theta} \cdot \mathbf{x}_- + b < 0$ for all data points. With this objective, the geometrical width between the two classes is given by $\frac{2}{\|\boldsymbol{\theta}\|}$. The objective now is to maximize the margin to obtain the optimal $\boldsymbol{\theta}^*$ and b^*.

In general, a larger margin indicates a lower generalization error for the classifier. The data points closest to the decision boundary are known as *support vectors*, and these data points have the greatest influence on the decision boundary. Vapnik proposed a maximum margin algorithm in 1963, and the SVM algorithm was introduced in 1992. The *Kernel trick* [51] allows the SVM to perform a powerful transformation on the data so that a non-linear decision boundary can be learned. To achieve this, the input data is transformed into a higher-dimensional space or Hilbert space, and a linear decision boundary is learned and transformed back to the original feature space. Gaussian kernels and polynomial kernels are some of the most commonly used kernels. Figure 2.8 shows the use of a Gaussian kernel to learn a nonlinear decision boundary.

Real-life data is rarely clean, and it can contain outliers. It is impossible to learn a decision boundary (whether linear or nonlinear) that can perfectly separate the two classes. Vapnik and Cortes also proposed a softmargin [52] approach by introducing the slack variables. These variables allow for the relaxation of the maximum margin constraint and allow either class outliers to be present on the incorrect side of the decision boundary. Different variations of SVM have also been proposed, including the least square SVM (LS-SVM) [53], one-class SVM for anomaly detection [54, 55], and adaptive SVM [56].

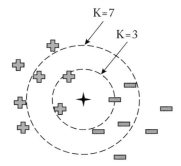

Figure 2.9: The K-nearest neighbor classification algorithm for two different values of K. When K is set to 3, the unlabeled data point is classified as positive. When K is set to 7, unlabeled data point is classified as negative.

2.3.3   THE K-NEAREST NEIGHBOR ALGORITHM

The k-Nearest Neighbors (KNN) [57, 58] algorithm is one of the simplest supervised machine learning algorithms. The KNN algorithm is an instance-based learning algorithm that does not learn a hypothesis function from the training data. Instead, the KNN algorithm stores all instances of the training dataset. The KNN algorithm can be explicitly used to classify input points to one of the possible discrete outcomes. The KNN algorithm works on the principle of neighborhood proximity. Neighborhood proximity is defined using a similarity metric, of which the distance-based metric is commonly used. A simple KNN model is shown in Figure 2.9. As seen in the figure, the KNN algorithm is nonprobabilistic and non-parametric in nature.

Given the training and test dataset, the KNN model's goal is to predict labels for the test dataset. The KNN model operates in the following manner: for the unlabeled data point (shown as a black star), the K most similar labeled data points are identified first (K closest labeled data points in this case), and classification is made from a simple majority vote of the K-nearest neighbors. The model performance is dependent only on the hyper-parameter value K. In Figure 2.9, for a K value of 3, the unlabeled data point is classified as positive. However, when K is set to 7, most neighbors are now from the negative class, and hence the unlabeled data point is classified as negative. Usually, a K value that has the best performance is selected. Also, models generated using larger values of K tend to generalize better. In some cases, the algorithm is run for multiple K values, and the decision is averaged. Another enhancement of the algorithm includes weighting the closest neighbors more when compared to the farther neighbors leading to the weighted nearest neighbor classification model.

Though the KNN algorithm is easy to use and intuitive, the biggest disadvantage is that KNN requires all the data points to be stored in the memory. Also, the KNN algorithm is computationally expensive since the distance must be calculated for every pair of data points. The complexity increases as dimensionality increases, and hence dimensionality reduction techniques are performed before using k-NN to avoid the effects of the *curse of dimensionality* [59].

The KNN algorithm can also be used for regression analysis [60] where the outcome of a dependent variable is predicted from the input independent variables. For the regression task, the predicted value is usually the mean (average) of the K-nearest neighbors.

2.3.4   NAIVE BAYES CLASSIFIERS

Naive Bayes classifiers [61] are simple probabilistic classifiers that are primarily based on the principle of Bayes' theorem. The term "Naive" indicates the strong assumption of conditional independence [62] for every pair of features (independent variables). That is, the algorithm assumes that all the d-dimensional input features $(\mathbf{x} : x_1, x_2, \ldots x_d)$ are independent of each other, and no correlation exists between them. Since it is a probabilistic model, naive Bayes' classifier outputs a posterior probability for a data point for each of the C classes. The class with the maximum posterior probability is chosen as the predicted class.

$$P(y_c \mid x_1, \ldots, x_d) = \frac{P(y_c)P(x_1, \ldots, x_d \mid y_c)}{P(x_1, \ldots, x_d)}; \quad c \leftarrow 1 \text{ to } C$$

Here, $P(\mid)$ indicates conditional probability and $P(,)$ indicates joint probability. With conditional independence, we have

$$P(x_i \mid y_c, x_1, \ldots, x_{i-1}, x_{i+1}, \ldots, x_d) = P(x_i \mid y_c)$$

Applying to all feature dimensions we have

$$P(y_c \mid x_1, \ldots, x_d) = \frac{P(y_c) \prod_{i=1}^{d} P(x_i \mid y_c)}{P(x_1, \ldots, x_d)}$$

The symbol $\prod_{i=1}^{d}$ indicates the product. Also, the denominator is a constant, which represents a joint probability distribution for all the input features, the prediction decision now becomes

$$P(y_c \mid x_1, \ldots, x_d) \propto P(y_c) \prod_{i=1}^{d} P(x_i \mid y_c) \tag{2.13}$$

$$\hat{y} = \arg\max_{y_c} P(y_c) \prod_{i=1}^{d} P(x_i \mid y_c); \quad c \leftarrow 1 \text{ to } C \tag{2.14}$$

Thus, the decision rule is to choose the class with the maximum probability; it is also known as *maximum a posteriori (MAP)*. Also, Equation (2.14) shows that decision is proportional to the maximum likelihood.

In practice, different Bayes' classifiers that differ mainly by the assumptions made on the data distribution $P(x_i \mid y_c)$ are used. For example, continuous data is typically assumed to be

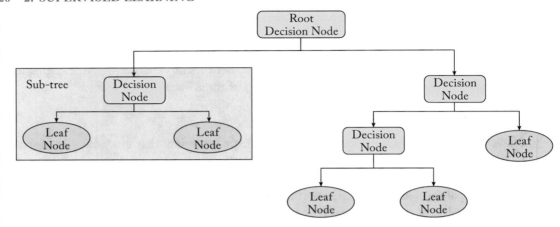

Figure 2.10: A representative decision tree showing decision nodes, branching, and leaf nodes.

distributed according to Gaussian distribution, and the corresponding classifier is termed Gaussian naive Bayes' classifier [63]. Similarly, multinomial naive Bayes' classifier [64] is associated with the multinomially distributed data.

2.3.5   DECISION TREES

Decision trees [65, 66] are nonparametric supervised learning models. A decision tree models a set of decision rules inferred from the input data features to predict the target outcome. Decision tree learns a hierarchy of "if-else" questions and uses them to predict the target variable. The deeper the hierarchical tree, the more complex the decision rules are. It is easy to implement and visualize the decision trees since they have a flow-chart like structure [67].

A representative decision tree is shown in Figure 2.10. Decision trees consist of (i) decision nodes, which answer a question, (ii) branches that represents the decision rule, and (iii) leaf nodes, which represent the outcome (or final decision). Thus, a complex decision rule is learned by recursively partitioning the tree. However, this can lead to over-fitting the training data. In practice, the tree's maximum depth is fixed before constructing it.

Induction and *Pruning* [68] are two important steps involved in constructing decision trees. In the induction step, the tree is constructed by selecting the best feature in the data to create the tree branch. The above process is performed recursively for all the features leading to many sub-trees. The process can be terminated when the entire feature set is accounted for or maximum tree depth is reached. During the decision tree pruning, redundant branches are removed from the sub-tree, which reduces the overfitting. Though decision trees require little or no data preparation (such as data normalization or scaling), outliers or noise in the data can lead to unstable decision trees.

2.4   SUMMARY

The supervised learning paradigm has access to ground truth or true labels in addition to the data. A supervised learning algorithm aims to learn a mapping function from input to the ground truth. Supervised learning can be again subdivided into regression and classification methods. Regression algorithms generally deal with predicting a continuous real-valued output whereas classification algorithm maps the input to discrete outcomes or categories. A linear regression model attempts to learn a linear mapping function, whereas polynomial regression models can learn a non-linear mapping function. Logistic regression is a classification algorithm that outputs class probabilities. Additional algorithms including SVM, K-nearest neighbor, Naïve Bayes, and decision trees were also described in this chapter. A support vector machine model produces the best decision boundary that maximizes the width of separation between classes while minimizing the generalization error. The K-nearest neighbor algorithm is a non-probabilistic instance-based algorithm that requires the training dataset to make an inference. A Naive Bayes classifier is a probabilistic model that assumes conditional independence on the input data. Decision trees are nonparametric models that infer (learn) decision rules from the input data.

CHAPTER 3

Unsupervised Learning

In unsupervised learning [69–71], the ML model does not have access to ground truth or any form of supervision. There are no explicit labels associated with the dataset. We only have access to $\mathcal{X}$, and $\mathcal{Y}$ is not available. The goal is often to extract the underlying structure and hidden knowledge in the data. Exploratory data analysis is often performed to find patterns and groupings. Unsupervised learning mainly consists of cluster analysis and often results in dimensionality reduction. Dimensionality reduction is often used in compression applications.

Cluster analysis [72, 73] deals with segmenting or partitioning the given data into distinct groups such that data points within a group are similar. In dimensionality reduction [74, 75], the goal is to reduce the number of feature dimensions (from d to p) in the data while preserving crucial information. The new representation of the data might often benefit an ML model by reducing the computational complexity. The data with dimensions reduced to two or three can also be visualized with ease. The general steps involved in an unsupervised learning algorithm is shown in Algorithm 3.2.

Algorithm 3.2 Unsupervised Learning Algorithm

Input: Training Dataset $\{\mathbf{x}^{(i)}\}_{i=1}^{N}$, Model
Output: Obtain Clusters C_{k} or Reduce dimensions ($d \to p$)
  1: **while** Model achieves required Performance **do**
  2:    Compute performance metric P for all $i \leftarrow 1$ to N
  3:    Update model parameters to improve P
  4: **end while**

3.1   COMMON CLUSTERING ALGORITHMS

In this section, we describe the basic ideas and properties of the most commonly used clustering algorithms in practice. The ML models can be classified into two styles based on their intrinsic learning mechanism.

Nonparametric models

In the nonparametric models [76, 77], there are no distinct assumptions on the distribution of the data or the hypothesis function. Models generally do not have any intrinsic parameters to optimize. The goal is to obtain a partitioning of given data such that the clusters are organized by their similarity. Data points that have high similarity should be part of the same clusters, and the dissimilar data should belong to different clusters. Nonparametric models should be used when there is no known prior information about the data. K-means algorithm is an example of nonparametric models.

Parametric models

In parametric models [78, 79], a strong assumption on the statistical distribution of the data is made. The ML model initially chooses a known hypothesis function but with unknown parameters. Then the parameters of the model are optimized using the data. The most common assumption is that the data comes from a family of normal distributions. Hence, the goal would be obtain the parameters *mean* and *standard deviation* for the different Gaussians. Gaussian mixture models (GMM) are an example of parametric models, and the parameters of the model are learned using the *Expectation-Maximization* algorithm.

3.1.1 THE K-MEANS ALGORITHM

The K-means clustering algorithm [80, 81] groups the given data into K number of clusters such that the data points within each cluster are similar to each other, and data points from different clusters are dissimilar. Similar to the k-NN algorithm introduced in Chapter 2, we use a similarity metric or a distance metric. Different distance metrics such as Euclidean, Mahalanobis, cosine, Minkowski, etc. are used. In general, the Euclidean distance metric is used more often.

The K-means algorithm is iterative and one of the simplest clustering algorithms. It clusters the data by separating them into K groups of equal variances, by minimizing within-cluster variances. Because there is no supervision or ground truth, it is rather harder to measure the performance of the algorithm. The K-means algorithm improves clustering efficiency by minimizing the within-cluster variances, also known as *inertia*, used as a cost function denoted by $J(C)$. However, the algorithm requires the number of clusters to be specified before running it.

Each observation or the data point is assigned to a cluster with the nearest mean, the mean represents the *Centroid* of that cluster. Thus, the K clusters can now be specified by the K centroids. For initialization, K data points are randomly chosen as K centroids. The algorithm's inner loop iterates over the following two steps to minimize the cost function $J(C)$:

1. Assign each observation $\mathbf{x}^{(i)}$ to the closest cluster centroid μ_j.

2. Update each cluster's centroid to the new mean of the points assigned to it.

$i \leftarrow 1$ to N Total of N observations or data points
$j \leftarrow 1$ to K Total of K clusters and hence K centroids

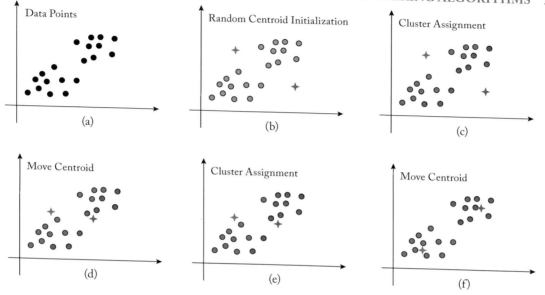

Figure 3.1: The K-means algorithm iteratively performs two important steps: (i) cluster assignment and (ii) move centroid.

The inertia or the within-cluster sum-of-squares is given by:

$$J(C) = \sum_{i=0}^{N} \min_{\mu_j \in C} \|\mathbf{x}^{(i)} - \mu_j\|^2 \tag{3.1}$$

$$J(C) = \sum_{k=1}^{K} \sum_{i=1}^{N} \|\mathbf{x}^{(i_C)} - \mu_k\|^2 \tag{3.2}$$

The iteration is stopped when the data points are no longer assigned to a different cluster or when no change is seen in the cluster centroids' location after each iteration. A critical initialization step is the definition of the value of K which needs to be specified before iterations start. Hence, K is often treated as a hyperparameter. Its value can be set if one is well aware of the data modalities. Otherwise, *elbow method* [83] is used to determine a proper K value that reasonably partitions the given data. Though the K-means algorithm is simple and fast, it often fails to perform accurate clustering when there is an overlap in the data. Also, the K-means algorithm is invariant to the data transformation, i.e., clustering results on data represented in Cartesian coordinates will differ from those of the polar coordinates [84]. In general, the algorithm performs poorly on nonlinear data as the algorithm is not globally optimal. We also note that there are hierarchical as well as genetic k-means clustering methods [85] where the goal is to create a hierarchy of partitions.

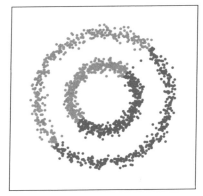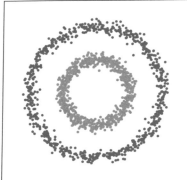

Figure 3.2: The results of running different clustering algorithms on circles dataset. The K-means clustering algorithm fails to consider the data manifold (left). Spectral clustering algorithm correctly identifies the clusters in the circles data (right).

3.1.2 SPECTRAL CLUSTERING

The spectral clustering algorithm [86, 87] overcomes some of the limitations of the K-means clustering algorithm. The spectral clustering algorithm has its root from spectral graph theory [88], which can readily be applied to a graph to obtain different communities. The algorithm can also be applied to non-graph data by explicitly constructing a graph. Given a set of N data points $x_1, x_2 \ldots x_N$, as a first step, a notion of similarity $s_{ij} \geq 0$ is defined between all pairs of data points x_i and x_j. A graph G with N vertices (or nodes) is constructed. Each vertex v_i in this graph represents a data point x_i. Two vertices are connected if the similarity s_{ij} between the corresponding data points x_i and x_j is positive or larger than a certain threshold, and the edge is weighted by s_{ij}. Once the graph is available, the Eigenvalues or *spectrum* information is obtained from the adjacency matrix, with the help of which clustering can be performed. The important steps involved in spectral clustering can be summarized as follows:

1. For given dataset $\mathcal{X}$ with N data points, construct a similarity graph G.

2. Compute the Laplacian matrix as $L = D - A$.

3. Find the number of zero Eigenvalues (say K).

4. Obtain the first K eigenvectors of its Laplacian matrix L as features for the data points

5. Run K-means on these features to separate objects into K clusters.

Spectral clustering easily overcomes the limitations of the K-means algorithm. Unlike the K-means algorithm, the spectral clustering results depend on the connectivity between the data points. Hence, it can perform well, even for nonlinear data. Finally, spectral clustering can be implemented using efficient linear algebra solvers [89].

3.1.3   GAUSSIAN MIXTURE MODELS

A Gaussian mixture model (GMM) [90, 91] is a parametric and probabilistic model that performs a soft-clustering for the given data points. Unlike the K-means algorithm, where a data point is strictly assigned to one of the K clusters, GMM provides the probabilities (or uncertainties) with which the data point belongs to each cluster. Finally, the cluster with maximum probability is chosen.

GMM is usually used when the given data is known to be multi-modal. The goal of a GMM is to represent a number of normally distributed subpopulations within an overall population. Moreover, each subpopulation is assumed to be from a Gaussian distribution. It is well known that a Gaussian distribution [92] can be completely represented by the mean (μ) and its standard-deviation (σ). A univariate Gaussian distribution, distribution with one-independent variable is represented as $x \sim \mathcal{N}(\mu, \sigma)$. Similarly, a multivariate Gaussian distribution is represented by the mean vector ($\boldsymbol{\mu}$) and the covariance matrix ($\boldsymbol{\Sigma}$) as $\mathbf{x} \sim \mathcal{N}(\boldsymbol{\mu}, \boldsymbol{\Sigma})$.

A Gaussian mixture model is a mixture of K different weighted Gaussians, where the mixture components are denoted by ϕ_k where $k \leftarrow (1, 2, \ldots K)$ such that $\sum_{k=1}^{K} \phi_k = 1$. The mixture component weights ϕ_k are also known as mixing probability, signifying how big or small each subpopulation is. The parameters mean $\boldsymbol{\mu}_k$ and covariance $\boldsymbol{\Sigma}_k$ for each Gaussian component is learned through the expectation-maximization algorithm. The expectation-maximization [93] algorithm is an iterative technique to obtain maximum-likelihood estimates for the given data. To obtain each of the model parameters and mixing probabilities that explain the data in the best possible way (*best-fit*).

$$p(\mathbf{x}) = \sum_{k=1}^{K} \phi_k \cdot \mathcal{N}(\mathbf{x}|\boldsymbol{\mu}_k, \boldsymbol{\Sigma}_k)$$

$$\mathcal{N}(\mathbf{x}|\boldsymbol{\mu}_k, \boldsymbol{\Sigma}_k) = \frac{1}{\sqrt{(2\pi)^d |\boldsymbol{\Sigma}_k|}} \exp\left(-\frac{1}{2}(\mathbf{x} - \boldsymbol{\mu}_k)^T \boldsymbol{\Sigma}_k^{-1}(\mathbf{x} - \boldsymbol{\mu}_k)\right)$$

Once we obtain the mixing probabilities and Gaussian parameters, the posterior probability that a data point belongs to the component C_k is computed using the Bayes' rule:

$$p(C_k|\mathbf{x}) = \frac{p(\mathbf{x}, C_k)}{p(\mathbf{x})} = \frac{p(C_k)p(\mathbf{x}|C_k)}{\sum_{j=1}^{K} p(C_k)p(\mathbf{x}|C_k)} = \frac{\phi_k \mathcal{N}(\mathbf{x}|\boldsymbol{\mu}_k, \boldsymbol{\Sigma}_k)}{\sum_{j=1}^{K} \phi_j \mathcal{N}(\mathbf{x}|\boldsymbol{\mu}_j, \boldsymbol{\Sigma}_j)}$$

Embedded Gaussian Mixture models have been used in conjunction with accelerometers in a study on ML for IoT applications [94].

3.2   FEATURE DIMENSIONALITY REDUCTION FOR UNSUPERVISED ML

Dimensionality reduction transforms data from a high-dimensional space (d) into a low-dimensional space (p) while ensuring that the low-dimensional data representation still retains

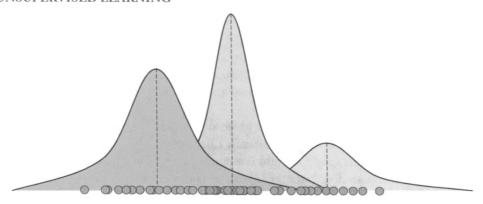

Figure 3.3: Gaussian mixture model showing three different Gaussian distributions.

most of the important information in the original data. By reducing the dimension of the original data, it becomes a lot easier to explore, visualize, and understand the relationships between the given features. In some cases, dimensionality reduction can help avoid over-fitting the given ML model by removing the inherent noise in the data. Dimensionality reduction can be achieved either through:

(i) **Feature selection**: The features are ranked according to their performance on a given task [95]. The feature subset with the highest rank is selected. Note that the selected subset of features is dependent on the given task and is often unreliable.

(ii) **Feature extraction**: The given feature space is transformed through statistical operations [96]. New independent features are obtained as a linear or nonlinear combination of the old features. This technique is often task-agnostic, and the transformed feature space is reliable.

3.2.1   PRINCIPAL COMPONENT ANALYSIS

Principal Component Analysis (PCA) [97, 98] is a linear dimensionality-reduction method that performs statistical transformations on the original dataset with a large set of independent variables to a smaller set. The transformed space is such that it retains most of the information from the original data. However, reducing the independent variables will inevitably lead to information loss. Usually, there is a trade-off between information loss and the number of dimensions. Reducing the number of dimensions will decrease the computational complexity for many ML models, but the accuracy of such models may be reduced.

One can think of PCA as a method that projects the given data points onto a reduced dimension while keeping the maximum information about the original data points in their reduced counterparts. In this case, by maximum information, we mean that the Euclidean distance

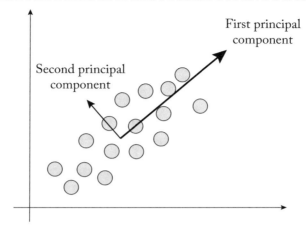

Figure 3.4: Principal Component Analysis: all the data points are projected along the diagonal direction. Most of the information (variance) in the data is retained when projected on first principal component.

between the original data points and their projected counterparts is minimal. Hence, data points that were "close" in the given feature space will produce lower-dimensional vectors that are still close to each other. The projection basis is the principal components obtained from the given data. Furthermore, the principal components are constructed such that the first principal component accounts for the largest possible variance in the data set. The common steps involved in performing PCA are as follows.

1. Mean normalize the given data ($\mathbf{X} \in \mathbb{R}^d$) such that new data ($\tilde{\mathbf{X}}$) has a mean of 0.

2. Compute the covariance matrix of mean normalized data ($\mathbf{\Sigma} = \tilde{\mathbf{X}}^T \tilde{\mathbf{X}}$).

3. Compute the eigenvectors and the eigenvalues of the covariance matrix by solving Eigenvalue decomposition ($\mathbf{VUV}^T = \mathbf{\Sigma}$).

4. Project the given data points onto the first p Eigenvectors by multiplying the first p Eigenvectors (also known as projection vectors) by the mean normalized data ($\mathbf{X}_{\text{reduced}} = \tilde{\mathbf{X}}\mathbf{V}[:p]$).

3.2.2   INDEPENDENT COMPONENT ANALYSIS

Independent component analysis (ICA) [99, 100] is another linear dimensionality reduction technique. ICA assumes that the given data points come from a mixture of independent components (and non-Gaussian). The goal is to find and separate the independent components (sub-components in the data). With prior knowledge, ICA works much better compared to PCA. For example, if it is known that many independent processes generate the data, it is easier to apply ICA to decompose the given data into its sub-components. *Cocktail-party problem* is a well-known example application of ICA [101]. In this scenario, two people (s_1, s_2) speak simultaneously in a room, and their speeches are recorded by two microphones (x_1, x_2) in separate locations. The two microphones record speeches from both speakers

$$x_1 = a_{11}s_1 + a_{12}s_2 \qquad\qquad (3.3)$$

$$x_2 = a_{21}s_1 + a_{22}s_2 \qquad\qquad (3.4)$$

$$x = As \quad \text{(matrix form)} \qquad\qquad (3.5)$$

Matrix A is called the mixing matrix, and its parameters depend on the distances of the microphones from the two speakers. Both A and s are generally unknown. The goal of the ICA is to recover s from x by finding A^{-1}.

In general, for the given data $\mathbf{X} \in \mathbb{R}^d$, the goal of ICA is to find p projection directions $(\mathbf{v}_1, \mathbf{v}_2, \ldots \mathbf{v}_p)$ such that the data projected on to these directions have maximum statistical independence. Note that both PCA and ICA perform statistical transformations on the data. While PCA performs up to second-order statistical transformations, ICA can perform even higher-order transformations. Also, in PCA, projection vectors are orthogonal, but ICA vectors are not orthogonal.

3.2.3 t-DISTRIBUTED STOCHASTIC NEIGHBOR EMBEDDING

t-Distributed Stochastic Neighbor Embedding (t-SNE) [102, 103] method is a nonlinear dimensionality reduction technique [75]. t-SNE is particularly employed to visualize higher dimensional data in lower-dimensional space. t-SNE embeds the given data points by preserving the small pairwise distances or local similarities. A similarity measure between pairs of data points in the original feature space (high dimension) and the embedding feature space (low dimension) is calculated. The above two similarity measures are optimized with the help of a cost function. The working of the t-SNE is explained in the following steps:

(i) A probability distribution $P(\cdot)$ is constructed over all pairs of points in the original feature space such that similar (nearby) data points are assigned a higher probability and dissimilar data points are assigned a lower probability.

(ii) A similar probability distribution $Q(\cdot)$ is constructed in the embedding space. A Student t-distribution with one degree of freedom is used to model both the distributions.

(iii) To make the two distributions similar, the Kullback–Liebler divergence (KL) [104] measure is employed. KL divergence measures the difference between the probability distributions [105, 106] ($P(\cdot)$ and $Q(\cdot)$) of the two-dimensional spaces.

Figure 3.5: t-Distributed Stochastic Neighbor Embedding (t-SNE) dimensionality reduction technique applied to handwritten digits.

Difference between t-SNE and PCA

The t-SNE is a probabilistic approach and is a nonlinear method. Hence, it is impossible to recover the original features from the embedding space. Unlike PCA, embedding feature space in the t-SNE cannot be expressed as a combination of input feature space. Also, PCA features maximize the variance in the given data by considering the overall pairwise distances, whereas the t-SNE method preserves the small pairwise distances or local similarities. The t-SNE algorithm is computationally expensive compared to the linear PCA method.

3.3   SUMMARY

Unsupervised learning deals with finding a structure or hidden pattern in the data. There is no explicit supervision provided with the data. The goal of the clustering algorithms is to find segmentation or partition of the data such that similar data items are grouped together. While K-means and spectral clustering perform hard-clustering, Gaussian mixture models perform soft-clustering. PCA is a linear dimensionality reduction technique that performs statistical transformations on the data. ICA is similar to PCA, but it works on the fundamental assumption that the data come from a mixture of independent components. t-SNE is a relatively recent approach that performs nonlinear dimensionality reduction.

CHAPTER 4

Semi-Supervised Learning

4.1 INTRODUCTION TO SEMI-SUPERVISED LEARNING

Semi-supervised learning [107–109] is one of the machine learning paradigms that typically uses unlabeled data and a limited amount of labeled data. Semi-supervised learning (SSL) falls in between supervised learning and unsupervised learning. When unlabeled data is used in conjunction with even a small amount of labeled data, it can produce a considerable improvement in accuracy over unsupervised learning. For example, consider Figure 4.1, a decision boundary of a classifier can be greatly improved when trained in conjunction with unlabeled data points.

Furthermore, labeling massive amounts of data for supervised learning is often very expensive and time-consuming. For example,

- In computer vision problems [110], manual annotations of object landmarks are costly. Expert annotators often identify and label the key points or object structures.

- In speech recognition [111], accurate transcription by a human expert is required, and it can take several hours to transcribe even an hour of speech.

- In web content classification [107], millions of web pages exist on the internet, and it often requires human intervention to classify the content.

On the other hand, unlabeled data is easy to collect. For example, many images with unlabeled object landmarks are available and speech utterances can be collected from pod-casts and videos. Semi-supervised learning is promising because it can utilize both labeled and unlabeled data to achieve better performance. From a different perspective, semi-supervised learning may achieve the same performance level as supervised learning, but with fewer labeled data samples. Semi-supervised learning algorithms often make assumptions about the data's underlying structure and use them as constraints in the learning problem. Common assumptions include:

- Continuity assumption: Points that are close to each other are more likely to share a label.

- Cluster assumption: When data form clusters, points in the same cluster are more likely to share labels.

- Manifold assumption: Data points in high dimension often lie on a manifold of a lower dimension.

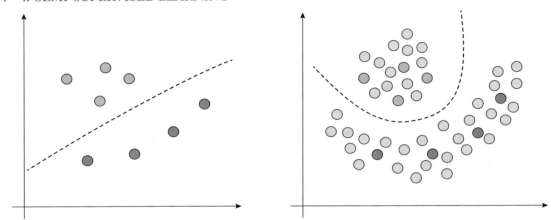

Figure 4.1: Semi-supervised learning utilizes a large volume of unlabeled data in conjunction with labeled data and has been shown to yield improved accuracy.

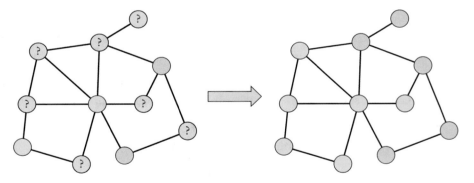

Figure 4.2: Given the graph structure and labels for a few nodes, the goal is to predict labels for the rest of the nodes in the graph.

This chapter focuses on popular semi-supervised learning models that typically use graphs to represent the input domain structure.

4.2   GRAPH-BASED SEMI-SUPERVISED LEARNING

In graph-based semi-supervised learning [112], the assumption is that both the labeled and unlabeled data are embedded in a low-dimensional manifold that may be reasonably expressed by a graph. The labels for unlabeled nodes are predicted utilizing the graph structure and the labeled nodes, as shown in Figure 4.2. If the data is not available as a graph, an explicit graph is constructed using a similarity score or measure. The simplest definition of a graph is "a collection of items connected by edges." In graph-based machine learning, each data instance is represented by a node/vertex in the graph. Nodes (or data instances) are connected with each other by an edge if they are similar. Thus, the edges of the graph provide a measure of similarity between all pairs of the data points.

We represent an unweighted graph using the tuple $G = (\mathcal{V}, \mathcal{E})$, where $\mathcal{V} = \{v_1, v_2, \ldots, v_N\}$ denotes the set of nodes with cardinality $|\mathcal{V}| = N$, $\mathcal{E}$ denotes the set of edges and $\mathcal{E} \subseteq \mathcal{V} \times \mathcal{V}$. The edges in the graph may be alternately represented using an adjacency matrix $\mathbf{A} \in \mathbb{R}^{N \times N}$. In addition, each node v_i may be endowed with a d-dimensional node attribute vector $\mathbf{x}_i \in \mathbb{R}^d$. Node attributes are also referred to a *graph signals* [113, 114]. We use the matrix $\mathbf{X} \in \mathbb{R}^{N \times d}$ to denote the features from all nodes.

Graphs offer a distinct advantage in modeling many of the real-world relational data. Graphs can inherently model the rich relational information between several entities and can incorporate complex pairwise relationships between them.

In scenarios where data is not available as a graph structure, a graph is constructed from the available data. The data is often assumed to be independent and identically distributed (*i.i.d*). Graph construction is performed without any supervision. i.e., label information is not required for constructing the graphs. An appropriate similarity metric that incorporates the domain knowledge is used to construct a graph from data instances. Two common construction methods are the k-nearest neighbor method and the ϵ neighborhood method. In the k-nearest neighbor method, for a given node, the connection is established only to the k-nearest neighbors, where the closeness is quantitatively defined using a similarity metric (for example, Gaussian kernel) or distance function (for example, Euclidean distance). In ϵ- neighborhood-based graph construction, an undirected edge between two nodes is added if the distance between them is smaller than ϵ, where $\epsilon > 0$ is a predefined constant.

Graph-based semi-supervised algorithms have become popular recently due to their state-of-the-art performance in many fields, including computer vision, speech processing, and natural language processing. Some of the classical problems that are addressed in graph-based machine learning include:

- Node Classification (Figure 4.3a)—Using labels of few nodes, predict the labels of other nodes using the graph structure and the labeled nodes [115].

- Link Prediction (Figure 4.3b)—Predict whether two nodes will establish a connection given the current graph structure [116, 117].

- Community Detection—Graph Clustering, partitioning nodes into clusters such that nodes in the same cluster have common properties [118].

- Graph Sampling—Learning a representative sample from a large graph [119].

- Graph Similarity—Identify if given two or more graphs similar structure (sub-networks).

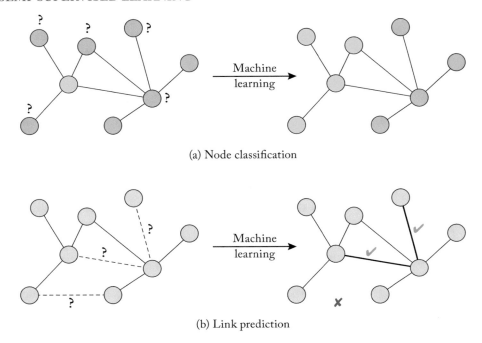

(a) Node classification

(b) Link prediction

Figure 4.3: Real-life applications in graph-based machine learning. The top figure shows a node classification problem, where the objective is to predict all the node labels given the labels of a few nodes. The bottom figure shows a link prediction problem. Given an incomplete graph with few links missing, the goal is to predict those missing links.

• Influential node selection—Identifying important nodes in the given graph [120].

Despite the variability in formulations employed to address the aforementioned problems, a recurring idea that appears in almost all of the approaches is to obtain embeddings for nodes in a graph termed as *node embeddings* [121], prior to carrying out the downstream learning task. This is the fundamental objective in graph-based machine learning, i.e., to learn to represent or encode the graph structure into node embeddings so that the downstream task-specific machine learning models can exploit it. An example of such an embedding algorithm on Zachary's karate club graph [122] is shown in Figure 4.4.

The fundamental premise of node embedding is to convert the discrete graph structure into continuous vectors (node embeddings). In the simplest form, the adjacency matrix indicating the connectivities can be treated as naïve embeddings for the nodes. However, it is well known that such cursed, high-dimensional representations can be ineffective for subsequent learning. Hence, there has been a long-standing interest in constructing low-dimensional embeddings that can best represent the network topology. However, the main challenge in graph-based ma-

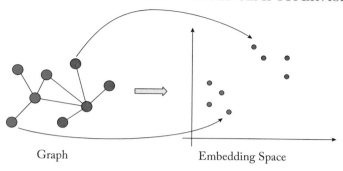

Graph   Embedding Space

Figure 4.4: Learning node embeddings: Random walk-based node embedding technique applied on Zachary's karate club graph. Nodes in the graph have distinct labels [left], and the learned 2-dimensional node embeddings preserve the community structure [right].

chine learning is that the structure of the graph/network is irregular when compared with images, or audio, or text. Images, speech, and other time-series data are structured by Euclidean grids. On the other hand, graphs are non-Euclidean; simple operations like convolution, translation, and downsampling cannot be explicitly defined on graphs.

4.2.1   MATRIX FACTORIZATION APPROACHES

Traditional graph-based machine learning approaches make use of graph Laplacian structure, where the graph Laplacian is defined as $\mathbf{L} = \mathbf{D} - \mathbf{A}$ and $\mathbf{D}$ is the degree matrix. One can perform Eigenvalue decomposition of the Laplacian matrix to obtain the Eigenvectors (also known as *spectrum* of graph). The low-dimensional latent representations or node embeddings are now chosen as the first p Eigenvectors. Label propagation [124] is another traditional approach that involves Laplacian regularization [125]. The Laplacian regularization method defines a real-values function $f(\cdot)$ on every node. The key assumption in this approach is that nearby nodes in a graph are likely to share the same labels. In addition to the cross-entropy loss defined on the labeled nodes, another cost function that minimizes the difference of the function values for every connected neighbor (all the edges) is employed (Equation (4.2)). Finally, gradient descent algorithm is used to minimize the Equation (4.1)

$$\mathcal{L}_{Total} = \mathcal{L}(y_{Labeled}, \hat{y}_{Labeled}) + \lambda \, \mathcal{L}(\mathbf{L}) \tag{4.1}$$

$$\mathcal{L}(\mathbf{L}) = \sum_{(i,j) \in \mathcal{E}} \| f(v_i) - f(v_j) \|^2 \tag{4.2}$$

The other graph-based machine learning methods that involve matrix decomposition include Spectral Clustering [87], stochastic factorization of the adjacency matrix [126], decomposition of the modularity matrix [127, 128], etc.

4.2.2   RANDOM WALK-BASED APPROACHES

An alternate class of approaches utilize the distributional hypothesis, popularly adopted in language modeling [129], where co-occurrence of two nodes in short random walks implies a strong notion of semantic similarity to construct embeddings—examples include *DeepWalk* [123] and *Node2Vec* [130]. The central idea of the DeepWalk, Node2Vec, or other random walk-based algorithm is to use short streams of randomly generated walks to define the notion of context for each node. Random walk-based methods have had long-standing success in quantifying similarities between entities in graph-structured data. Formally, a random walk is a stochastic process with a set of random variables defined as nodes chosen at random from the neighbors of each vertex in the sequence. The algorithm starts with a node at random, say v_i, which is the current node, then randomly selects a node v_j from the neighbors of v_i. Now, v_j is the current node, and this process is repeated till the desired walk length is reached. The random walk's ability to reveal the local structure makes it a natural tool for extracting information from graphs. Once several random walks are generated for each node, a skip-gram model or a hierarchical softmax is employed to generate dense embeddings.

4.2.3   GRAPH NEURAL NETWORKS

The unprecedented success of deep learning with data defined on regular domains, e.g., images and speech, has motivated its extension to arbitrarily structured graphs. While the aforementioned approaches effectively preserve network structure, semi-supervised learning with graph-structured data requires feature learning from node attributes to effectively propagate labels to unlabeled nodes. Since convolutional neural networks (CNNs) have been the mainstay for feature learning with data defined on regular grids, the natural idea is to generalize convolutions to graphs. Existing work on this generalization can be categorized into *spectral* approaches [132, 326], which operate on an explicit spectral representation of the graphs, and *non-spectral* approaches that define convolutions directly on the graphs using spatial neighborhoods [327, 330]. More recently, *attention* models [331] have been introduced as an effective alternative for graph data modeling, which mainly relies on attention mechanisms for feature learning.

Graph Convolutional Networks (GCNs) [131, 132] employ a message passing framework, where the messages from the neighborhood are aggregated. GCNs operate by weight sharing, just like in CNNs. The aggregated node features are transformed using the learnable weight, and an additional non-linearity is applied to these transformed features. The blue node in Figure 4.5 updates its features by (i) first aggregating the information from the neighboring nodes, (ii) transforming the aggregated features by learnable weight matrix $\mathbf{W}$, and (iii) applying an optional nonlinearity. The message passing mechanism in the GCN consists of two important functions.

1. Message Function (M): The message for the node v_i is obtained as m_i using the message function M. In the equation below, $\mathbf{h}_i$ and $\mathbf{h}_j$ indicate the features of the nodes v_i and v_j,

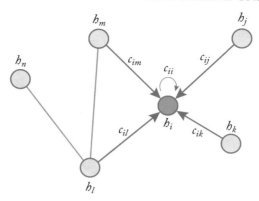

Figure 4.5: Graph Convolutional Networks employ a message passing framework that allows neighborhood feature aggregation and non-linear transformation.

respectively, and e_{ij} denotes the weight of the edge connecting the nodes v_i and v_j.

$$m_i = \sum_{j \in \mathcal{N}_i} M(\mathbf{h}_i, \mathbf{h}_j, e_{ij})$$

2. Update Function (U): The aggregated messages are received by node v_i. Usually, a simple mean or sum update function is used.

$$\mathbf{h}_i^+ = U(\mathbf{h}_i, m_i)$$

The two operations involved in message passing is summarized in the Equation (4.3), and c_{ij} denotes the normalization constant.

$$\mathbf{h}_i^+ = \sigma \left(\sum_{j \in \mathcal{N}_i} c_{ij} \mathbf{h}_j \mathbf{W} \right) \tag{4.3}$$

$$c_{ij} = \frac{1}{\sqrt{d_i d_j}} \tag{4.4}$$

4.3 POSITIVE-UNLABELED LEARNING

Positive-unlabeled (PU) learning [133] is a sub-category in semi-supervised learning in which the labeled data points are from the positive class only. The fundamental assumption in PU learning is that the unlabeled data points can contain both positive and negative examples. Such scenarios are common in medical diagnosis and knowledge base completion. One naive approach

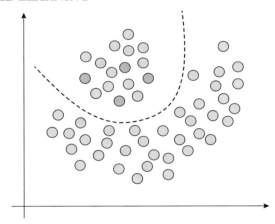

Figure 4.6: Positive-unlabeled learning: Data points in green belong to positive class. The gray data points could belong to either positive or negative class. The goal is to learn a classifier given only a few positively labeled data points.

to solve PU learning is identifying a set of unlabeled points that can be confidently labeled as negatives. The authors in [134] refer to these data points as *reliable negatives*. Reliable negatives can be identified by using a similarity metric and comparing them with the known positive labels. Once there are enough positive and negative data points, it is easier to train a standard classifier. The trained model can then be used to evaluate the remaining unlabeled data points. This method is commonly known as *two-step approach*.

Another approach in PU learning applies weights to the unlabeled data points. The weights represent the likelihood of the data point belonging to a positive or negative class [135]. A non-negative risk estimator is proposed in [136] to make the classifier model more robust against overfitting. Recently, bagging or bootstrap aggregating has become a popular approach to improving PU learning performance. Please refer to [137, 138] for a detailed survey of algorithms and applications on PU learning. A theoretical comparison of PU learning against the standard positive-negative learning is given in [139]. Some of the applications of PU learning can be seen in biomedical research such as disease gene identification [140], online data stream classification [141], and energy load forecasting [142].

4.4   SUMMARY

Semi-supervised learning uses a large amount of unlabeled data along with a few limited labeled data. This produces a considerable improvement in learning accuracy compared to unsupervised learning, but without the need for complete supervision. Also, labeling data is an expensive operation, and expert help is often required. SSL algorithms often make use of a graph to model the data domain. The fundamental assumption is that both the labeled data and unlabeled data are embedded in a low-dimensional manifold, and it may be expressed as a graph. One common approach in graph-based SSL is to learn node embeddings from the labels and graph structure. Node embeddings can facilitate downstream machine learning tasks. Matrix factorization and random walk-based methods are popular examples for embedding methods. Positive unlabeled learning is a scenario where only a few positive labels are available, and the unlabeled data can come from both positive and negative classes.

CHAPTER 5

Neural Networks and Deep Learning

In this chapter, a brief introduction to the field of artificial neural networks is provided with a focus on deep learning [9], neural network training, and different architectures. Artificial neural networks are powerful pattern recognition machines, and they have proved to be the most successful. Neural networks and deep learning are quite successful at end-to-end learning, and they do not require feature engineering as in traditional machine learning techniques such as SVM and decision trees. Deep learning has achieved unprecedented results in various fields, including signal processing [143, 144], computer vision [145], speech and audio processing [146], and natural language processing [147, 148].

5.1 PERCEPTRON: BASIC UNIT

The most basic unit of artificial neural networks or any deep learning architecture is a *perceptron* or a *neuron* which is inspired by the biological neuron [17]. Figure 5.1 shows a simple neuron consisting of (i) weights, (ii) bias (or threshold), and (iii) nonlinearity (or activation function). As will be shown, a perceptron is capable of performing a binary classification for the given input. The given inputs are multiplied by the perceptron's weights, and the sum of the weighted input is obtained. The sum is then compared with a threshold or a bias, and finally a nonlinear activation function is applied. If the weighted sum is greater than the threshold, then the perceptron outputs a positive value and a negative value otherwise.

The activation function mentioned above is a step-function that is nonlinear and can easily perform binary classification. Many such activation functions are available in literature including sigmoid, tanh, and ReLU. The nonlinearity in the activation allows the neuron to learn complex decision boundaries. Note that the perceptron's weights, including the bias, are learned during training. The perceptron is trained using a supervised learning Algorithm 2.1. During training, the perceptron can identify important input features, and they will be weighted more than other input features. Though the perceptron is very limited in terms of its learning power, by comprising several layers of stacked neurons, one can obtain more powerful artificial neural networks.

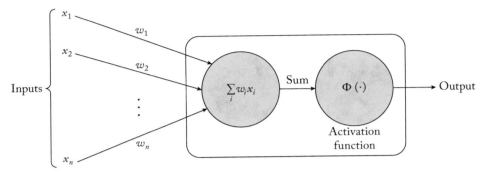

Figure 5.1: Perceptron/Neuron: a basic unit of an artificial neural network.

5.2   MULTI-LAYER PERCEPTRON

Artificial Neural Networks, also known as *multi-layer perceptrons* (MLP) [149], are obtained by repeatedly stacking several layers of neurons. The neurons in the current layer receive the outputs of the neurons from the previous layer as its inputs. Similarly, the current layer's neuronal output is fed as input to the next layer, as shown in Figure 5.2. The final layer is the output layer, a simple softmax classifier that outputs unnormalized log probabilities. In a nutshell, an MLP consists of three important layers: (i) input layer: where the input data is provided; (ii) hidden layers: multiple layers of stacked neurons, which allows the MLP to learn more sophisticated and complex patterns from the data; and (iii) output layer: outputs either class probability for classification problems or a continuous value for regression problems.

The term *deep learning* refers to several layers of stacked neurons that allow MLP to learn multiple representations of the input data. Each layer learns an abstract representation of the data that is generally different from the previous layers. The number of hidden layers employed in an MLP often depends on the complexity of the data, and the number of layers can vary from a few tens to thousands.

Information flows through the neural networks in two ways: (i) forward propagation and (ii) backpropagation. The data flow from the input layer to the output layers with multiple transformations from the hidden layers in forward propagation. In the output layer, the MLP model makes a prediction for the given input data. In backpropagation, the MLP model adjusts its neuron parameters (or weights) considering its prediction error. The activation function used in each neuron allows MLP to learn a complex function mapping. The MLP architecture used for handwritten digit classification is shown in Figure 5.2. Input to the model is flattened image $\mathbf{x}$ of dimension $784(28 \times 28)$, the output of the first and second hidden layer is given by

$$\mathbf{h}_1 = \sigma(\mathbf{W}_1\mathbf{x} + b_1)$$

$$\mathbf{h}_2 = \sigma(\mathbf{W}_2\mathbf{h}_1 + b_2)$$

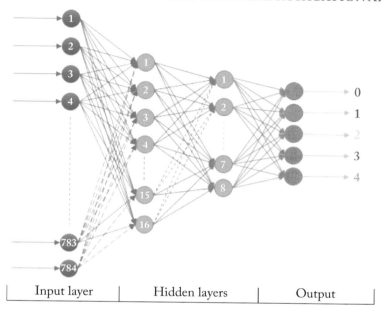

Figure 5.2: An artificial neural network with two hidden layers with 16 and 8 neurons, respectively. Input to the neural network is the resized spectrogram image and output of the neural network is the probability for each class.

Where $\mathbf{W}_1$ of size (16×784) is the weight matrix from input to first hidden layer and $\mathbf{W}_2$ of size (8×16) is the weight matrix from first hidden layer to second hidden layer. b_1 and b_2 are biases of size 16 and 8, respectively. Finally, the output class of the MLP is obtained as:

$$\hat{y} = argmax\left(\phi_{softmax}(h_2)\right)$$

5.3   TRAINING USING THE BACKPROPAGATION ALGORITHM

Though it is straightforward to train a single perceptron, a more sophisticated algorithm is required to train the whole neural network. This is because the output of the MLP not only depends on the neurons in the final layer but also depends on all the neurons from the input layer to the hidden layer and all the way to the output layer. Thus, the gradient of the loss function needs to be computed with respect to the weights of the neurons in the final layer, hidden layer, and the input layer as well. Recall that the gradient provides information about the rate of change of a function with respect to given variable. This technique is commonly known as *gradient descent* [150]. Intuitively, this is equivalent to finding the maximum slope while climbing a mountain at the current position. In this scenario, the mountain represents the cost function

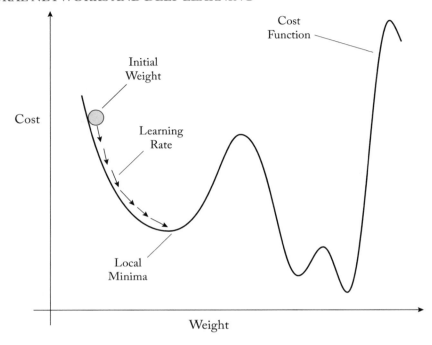

Figure 5.3: A gradient descent technique is used to update the model parameters with the aim to reduce the loss.

or the loss function. We go down the mountain (instead of climbing up) in the direction of the maximum slope by taking a small step since we are interested in reducing the loss. An intuitive figure for the same is shown in Figure 5.3.

For a single neuron with parameter $\boldsymbol{\theta}$ and loss function L, the gradient is given as $\frac{\partial L}{\partial \boldsymbol{\theta}}$. And the update rule is given by

$$\boldsymbol{\theta}^+ = \boldsymbol{\theta} - \alpha \cdot \frac{\partial L}{\partial \boldsymbol{\theta}} \tag{5.1}$$

Here α is the *learning-rate* that defines the rate of convergence to the local-minimum. From a different perspective, gradient descent is an optimization tool used to reduce the loss or the prediction error.

Backpropagation, as the name suggests, allows for the backward propagation of the errors [151, 152]. For the MLP in Figure 5.2, model parameters $(\mathbf{W}_1, \mathbf{W}_2)$ need to be updated during training using the gradient descent. As mentioned earlier, we need to compute the gradient of the loss with respect to all layers' parameters. It is easier to compute the gradient $\frac{\partial L}{\partial \mathbf{W}_2}$ and to compute the gradient $\frac{\partial L}{\partial \mathbf{W}_1}$ we need to propagate the error backward using the chain rule from calculus. The gradient $\frac{\partial L}{\partial \mathbf{W}_1}$ is computed by first computing the gradient $\frac{\partial L}{\partial \mathbf{h}_1}$ and then $\frac{\partial \mathbf{h}_1}{\partial \mathbf{W}_1}$.

$$\frac{\partial L}{\partial \mathbf{W}_1} = \frac{\partial L}{\partial \mathbf{h}_1} \cdot \frac{\partial \mathbf{h}_1}{\partial \mathbf{W}_1} \tag{5.2}$$

For a neural network with more than two hidden layers, the gradients of the parameters are computed by recursively applying the chain rule given by Equation (5.2). Once the gradient for all the parameters is obtained, the parameters are updated using the gradient descent rule given Equation (5.1). The complete backpropagation algorithm is summarized in Algorithm 5.3. The weights of the neural networks are initialized randomly to begin the training.

Algorithm 5.3 Backpropagation Algorithm for Training Neural Networks

---

Input: Training Dataset $\{\mathbf{x}^{(i)}, y^{(i)}\}_{i=1}^{N}$ and MLP Model parameters $\mathbf{W}_1, \mathbf{W}_2, \cdots \mathbf{W}_P$
Output: Trained Model parameters $\mathbf{W}_1^*, \mathbf{W}_2^*, \cdots \mathbf{W}_P^*$
 1: **while** Model achieves required Performance **do**
 2:     Perform forward propagation to compute predictions $\hat{y}$
 3:     Compute loss $L(y^{(i)}, \hat{y}^{(i)}); i \leftarrow 1$ to N
 4:     Compute gradients by applying chain rule $\frac{\partial L}{\partial \mathbf{W}_P}, \frac{\partial L}{\partial \mathbf{W}_{P-1}}, \cdots, \frac{\partial L}{\partial \mathbf{W}_2}, \frac{\partial L}{\partial \mathbf{W}_1}$
 5:     Update $\mathbf{W}_j$ to minimize $L; j \leftarrow 1$ to P
        $\mathbf{W}_j^+ = \mathbf{W}_j - \alpha \cdot \frac{\partial L}{\partial \mathbf{W}_j}$
 6: **end while**

5.4   ACTIVATION FUNCTIONS

In this section, different activation functions [153] used in deep learning are introduced. Every activation function takes an input and applies nonlinearity over it. Since the nonlinearity is a predetermined mathematical operation, it is possible to differentiate the function. This allows for backward propagation of the loss through the network. The nonlinearity applied over several layers allows the neural networks to learn a complex relationship that is generally not possible otherwise. Commonly used activation functions (Figure 5.4), along with their limitations, are given below.

1. **Sigmoid function**: The sigmoid function takes in an input and squashes it between 0 and 1. The output of a sigmoid is usually interpreted as a probability for this reason. Mathematically, it is given as

$$\sigma(\mathbf{z}) = \frac{1}{1 + e^{-\mathbf{z}}}$$

The sigmoid function is applied element-wise to its inputs. Some of the limitations are (i) saturated neurons kill the gradients (also known as *vanishing gradient* problem), (ii) outputs are not zero-centered, and (iii) computing exponential function is expensive.

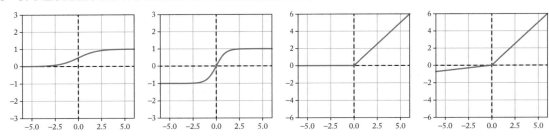

Figure 5.4: Activation functions employed in neural networks (from left to right): Sigmoid, Tanh, ReLU, Leaky ReLU.

2. **Tanh function**: Similar to sigmoid, the hyperbolic tangent function takes in a real-valued input and squashes it to a value between -1 and 1. Mathematically, it is given as

$$\tanh(\mathbf{z}) = \frac{e^{\mathbf{z}} - e^{-\mathbf{z}}}{e^{\mathbf{z}} + e^{-\mathbf{z}}}$$

Though it is zero-centered it still kills the gradients in the saturated regime.

3. **ReLU function**: ReLU is short for the rectified linear unit, and it is popularly used in modern deep learning models as it is known to converge at a faster rate. Mathematically, it computes the following function $f(\mathbf{z}) = \max(0, \mathbf{z})$. In other words, the activation function is a piece-wise linear function thresholded at 0. ReLU is very inexpensive to compute compared to sigmoid and tanh functions. The vanishing gradient problem is eliminated in the positive regime, but some neurons which output 0 can still die.

4. **Leaky Relu function**: Leaky ReLU function attempts to fix the dying neuron problems encountered in ReLU. In the negative data regime, instead of outputting a 0, it outputs a small value with a negative slope. For input $z < 0$, it outputs αz where α is a small number (usually 0.01 or 0.02).

5.5   NEURAL NETWORK REGULARIZATION: AVOIDING OVER-FITTING

Neural networks with several hidden layers, though they have strong representation power, can often overfit the given data, especially when noise (or outliers) is present. An example of the over-fitting scenario is shown in Figure 5.5. Note that if the neural network is underfitting the given data, one can increase the model's capacity by increasing the number of hidden layers or increasing the number of neurons in the hidden layers. Other challenges in training artificial neural networks include initializing the network parameters, overfitting, and long training time. There are various techniques to address the aforementioned problems. Batch normalization [154], weight

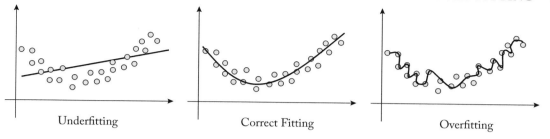

Figure 5.5: Neural networks with different solutions showing underfitting, correct fitting, and overfitting.

normalization [155], and layer normalization [156] all help in accelerating the training of deep neural networks. Weight regularization, dropouts [157] and data-augmentation [158] help in reducing overfitting. In this section, we will mention a few techniques that reduce over-fitting in the neural networks.

1. **Weight Regularization or L2 Regularization**: This is the most common form of regularization [159, 160] technique to avoid overfitting. In neural network weight regularization, a penalty term dependent on the weight parameters is added to the cost function or loss function. In L2 regularization, the penalty term is the sum of squares of all weights. Intuitively, L2 regularization penalizes the peaky weight vectors and instead prefers the diffusion of weight vectors encouraging the model to utilize all the input features to influence the output.

2. **Data Augmentation**: The performance of a neural network model with a very high capacity (many hidden-layers) continues to improve as more and more data is used to train it. Thus, a simple way to reduce overfitting for a high-capacity neural network is to increase the training data size [158]. However, it may not always be feasible to collect more data in critical applications (medical images, for example). In such scenarios, one can use the available data to generate more data through transformations such as shift, scale, flip, and random cropping. This basic idea is referred to as *data augmentation*.

3. **Dropout**: Dropout [157] is a relatively new technique of regularizing the neural network models. It is a simple and effective technique introduced by Srivastava et al. During training, dropout is implemented by only keeping a neuron active with some probability p, or setting it to zero otherwise as shown in Figure 5.6. The parameters of the neurons that are turned off are not updated. This forces other neurons to learn from the input data. Dropout can be considered as sampling a neural network so that neurons between the layers are not fully connected. Also, dropout is not applied during the inference.

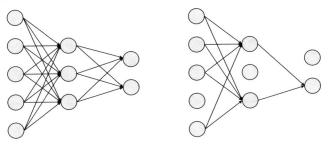

Neural Network   Neural Network with Dropout

Figure 5.6: Dropout turns off a neuron with the probability p. The left figure shows a fully connected neural network layers and right figure shows neural networks layers where some neurons are turned off.

4. **Ensemble Learning (Bagging)**: Ensemble learning [161] is a simple yet effective approach to combat overfitting in neural networks. The idea is to train multiple neural network models and combine their predictions. Since neural networks are trained in a stochastic manner, different weight parameters are obtained each time they are trained. Thus, neural networks can be seen as low bias and high variance models sensitive to input data. By combining the predictions from several neural network models, one can average out the high variance and reduce generalization error. This class of techniques is known as ensemble learning and is often used with other ML models.

In the next couple of sections, we will study different neural network architectures for various kinds of data.

5.6   CONVOLUTIONAL NEURAL NETWORKS

Convolutional Neural Networks (CNNs) [5, 162] are very similar to multi-layer perceptron conceptually. CNNs are predominantly made up of two important neural network layers: (i) convolutional layers and (ii) sub-sampling or max-pooling layer. To distinguish CNNs from MLP architecture, neural network layer introduced in Section 5.2 is commonly referred to as a *fully connected layer* or a *dot-product layer*. The convolutional layer is made up of neurons that have trainable (learnable) weights and biases. However, the neurons in the convolutional layer receive only a part of the input at a given time termed as the *receptive field*. In fact, this arrangement of neurons is inspired by the connectivity pattern in the visual cortex of the human brain.

CNNs are used on images since they explicitly capture the spatial information in the image. Suppose an ordinary, fully connected neural network is used on images. The images are flattened or vectorized first to be fed as inputs to an MLP model. By vectorizing the images, spatial dependencies, which are predominant in images, is lost. CNNs, on the other hand, take advantage of these spatial dependencies for learning a mapping function.

5.6.1   CONVOLUTIONAL LAYER

A convolutional layer processes volume of activations rather than a vector and produces feature maps. The convolutional layer consists of learnable filters (neurons, in other words) that have a small but specific spatial dimension compared to the image size. A filter is often represented by its height, width, and depth. The depth refers to the number of channels in the input image. Natural images are usually made of three primary channels: red, green, and blue. For efficient implementation, square filters are commonly employed. So a filter of dimension (f, f, d) has $f \times f \times d$ (plus 1 for bias) learnable parameters. This filter is convolved with the entire input to obtain an *activation map*. In other words, the filter slides over the image at every possible spatial location. A dot product is computed for the overlapping values, and a nonlinearity is applied. The entire operation is summarized in Figure 5.7.

The activation map can be interpreted as a re-representation of the given image. Akin to dot-product layers, multiple filters (neurons) are often used in convolutional layers. The filter preserves the local-connectivity, and the receptive-field for a filter is given by its size. Note that the same weights are used for a filter while convolving over the image.

5.6.2   MAX-POOLING LAYER

In addition to conv layers, max-pooling layers are commonly employed in CNNs. The pooling layer aims to gradually reduce the spatial dimension of the activation maps without losing the most prominent information. Thus, pooling layers help reduce the number of trainable parameters, allowing a faster computation without sacrificing the models' accuracy and reducing the overfitting. It is important to note that the pooling layers do not have any learnable parameters. They operate on a small region of activation map; consider a 2×2 window, for example. A max-pool operation takes four values as inputs and outputs the maximum input value. The pooling layer is applied to all the channels (or depth) in the activation maps independently. The operations are summarized in Figure 5.8

5.6.3   CONVNET ARCHITECTURE

Any ConvNet architecture consists of repeatedly stacked convolutional layers and max-pooling layers. In addition to these two, a fully connected layer is employed toward the end. A fully connected layer reduces the dimension and is often the final layer with which the output scores are computed. Thus, the input image undergoes various transformations by convolutional layer and max-pooling layer and produces an output score for each class-probability. All the network parameters (from both conv layers and fully-connected layers) are trained in an end-to-end fashion since the complete network is differentiable. The gradient for all the parameters is computed using backpropagation. A simple CNN architecture is shown in Figure 5.9. There are

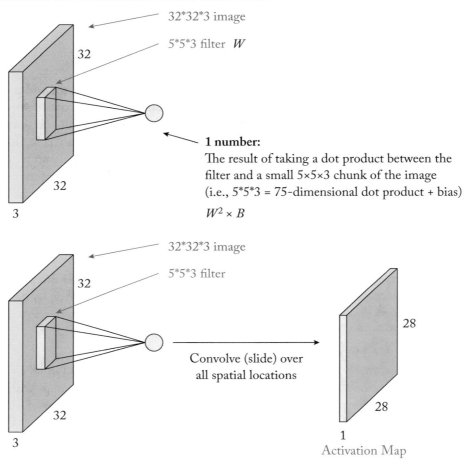

Figure 5.7: A convolutional layer consists of learnable filters which is used to convolve over the entire image space. An activation map is obtained as a result of convolving a filter over the input image volume.

several CNN architectures present in literature as datasets have grown larger and hardwares are becoming more powerful. CNN architectures mainly vary in the number of layers, filter sizes, etc. Some of the popular CNN architectures include LeNet [5], AlexNet [163], VGGNet [164], Inception [165], and ResNet [166].

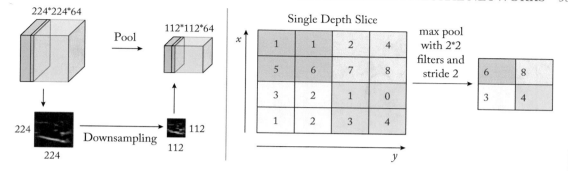

Figure 5.8: A max-pooling layer preserves the depth of the input activation map. The 2×2 windows reduces the spatial dimension of the activation map by half.

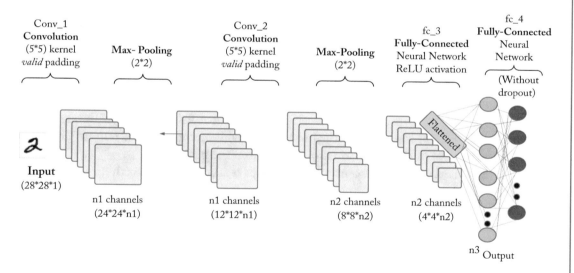

Figure 5.9: A ConvNet architecture consisting of two convolutional layers, two max-pooling layers, and two FC layers.

5.7   RECURRENT NEURAL NETWORKS

Recurrent Neural Networks (RNNs) [167, 168] are a type of artificial neural network architecture where the output depends on present inputs and outputs from the previous steps. RNNs are commonly used for sequential processing of data or time-series data where definite temporal dependencies exist. In a traditional neural network (fully connected or convolutional architecture), the usual assumption is that the inputs (also outputs) are independent of each other; however, this is not the case for applications such as natural language processing, weather prediction,

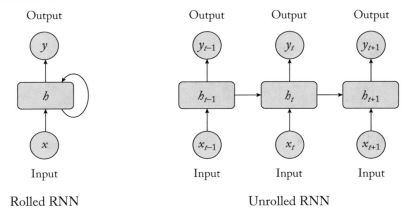

Figure 5.10: A recurrent neural network adds the previous state output as its current input. Left image shows an rolled RNN, and the right figure shows an unrolled RNN.

speech recognition, etc. A well-defined temporal dependency exit in these applications. For example, the next word in a sentence can be predicted by looking at all the words that came before. So a notion of "memory" is required in the neural network architecture. RNN achieves this notion of memory by incorporating a self-loop, as shown in Figure 5.10. Through this looping mechanism, RNN has access to the hidden state while computing the output. The hidden state is representative of all the information learnt from the past inputs.

Standard gradient descent techniques can not be applied directly to train RNN because of the looping mechanism. Instead, RNNs are trained using backpropagation through time (BPTT) algorithm [169]. Though theoretically, RNN can make use of the hidden state for an infinite number of steps, problems like vanishing gradients limit the step size to a finite number (short-term memory). Bidirectional RNNs [111] is a variant of RNNs in which the current output depends on the past and future inputs. The short-term memory problem in RNNs is overcome in the long short-term memory (LSTM) [170] architecture.

5.8   UNSUPERVISED REPRESENTATION LEARNING USING NEURAL NETS

The goal of representation learning [97] is to extract atomic features from the data. The learned representations can facilitate the transfer of knowledge across a myriad of tasks (also known as *multi-task learning*). In classical training of neural networks, the models are optimized against a specific task (image classification, for example). Instead, we can train the neural nets to learn good feature representation for the image data without any task. The choice of the representations should depend on the downstream tasks. In this section, we will introduce two neural network architectures that enable better representation learning.

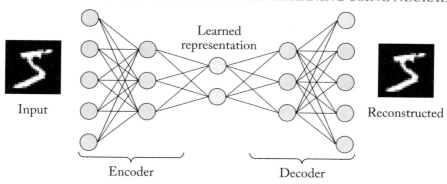

Figure 5.11: An auto-encoder architecture with encoder and decoder. The encoder compresses the given input image where as the decoder reconstructs the original image data from the learned representation.

5.8.1   AUTO-ENCODERS

Auto-encoders [171] are unsupervised representation learning techniques that employ neural networks. Auto-encoders specifically impose a bottleneck in their neural network architectures. This enables the neural nets to learn a compressed knowledge representation. Auto-encoder consists of two parts: (i) *Encoder*—The data x is transformed into a low-dimensional representation z by the encoder; and (ii) *Decoder*—Decoder reconstructs the input data $\hat{x}$ by transforming back the hidden representation (encoded data with bottleneck). In other words, auto-encoder is trying to learn an identity mapping function $\hat{x} = f_\theta(x)$. The neural network is then trained to minimize the error in the reconstruction of the input. A generic neural network-based auto-encoder is shown in Figure 5.11. A varying number of auto-encoder architectures exist in practice, such as sparse auto-encoders [172], denoising auto-encoders [173], and variational auto-encoders [20].

5.8.2   GENERATIVE ADVERSARIAL NETWORKS

Generative adversarial networks (GANs) [21, 174] are generative models (nonparametric) capable of learning underlying data distributions so that they can generate new data with the same statistics as the training data. GANs consists of two neural network models that are competing against each other: (i) *Generator* (G): Generator is a neural network model that generates new data and (ii) *Discriminator* (D): Another neural network model that classifies whether the given data is real or it was generated. The generator and discriminator are analogous to an art forger and art inspector, respectively. The generator is trained to generate more realistic looking data so that the discriminator can be fooled. In contrast, the discriminator is trained to distinguish between the generator's fake data and real data. A conceptual idea of a GAN is shown in Figure 5.12.

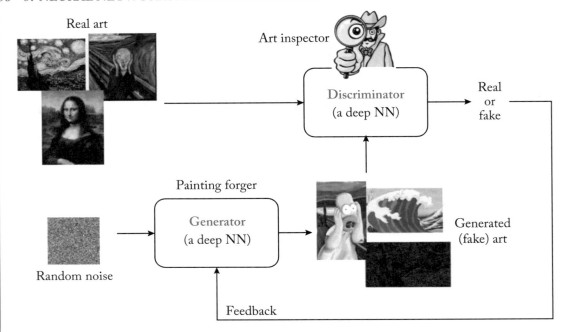

Figure 5.12: A generative adversarial network (GAN) with generator and discriminator components that are competing against each other throughout the training.

The discriminator in practice is a neural network classifier that outputs a 0 (for fake) or 1 (for real). The discriminator is trained with real-world data and the generators fake data. Using the ground truth, we compute the loss, and through backpropagation, we update the weights of the discriminator to improve its performance. On the other hand, the generator takes a random noise z as its input. The generator applies multiple nonlinear transformations to the noise vector z to generate realistic-looking data.

We use noise in order to generate a wide variety of realistic-looking data. Finally, the generated image is fed to the discriminator. The feedback from the discriminator allows the generator to update its weights so that it can fool the discriminator even better. Thus, the discriminator tries to minimize its loss by identifying the fake data correctly, whereas the generator tries to maximize the discriminator's loss by fooling the discriminator. Consequently, GANs suffer from training instabilities since both G and D have to be trained simultaneously. Training instability can arise when one component D outperforms G or vice versa. Many techniques are recently proposed to stabilize GAN training [175, 176]. Other variants of GANs include conditional GAN [177], controllable GAN [178], InfoGAN [179], styleGAN [180].

5.9 SUMMARY

Artificial neural networks (ANNs) are made up of repeatedly stacked layers of neurons or perceptrons. A perceptron is a basic unit in a neural network capable of performing binary classification. Each layer in an ANN learns an abstract concept different from the previous layer. ANNs use the backpropagation algorithm to modify the weights and optimize the model to minimize prediction errors. ANNs also employ activation functions to perform nonlinear transformations on the input. The nonlinearity applied over several layers allows the neural networks to learn a complex relationship that is impossible otherwise. Neural networks as such may be prone to overfitting, and several approaches exist in practice to avoid overfitting. The convolutional neural network architecture works exceptionally well with spatial data such as images, whereas recurrent neural networks generally work well on temporal or time-series data. Auto-encoders and generative adversarial networks are applications of neural networks for unsupervised representation learning.

CHAPTER 6

Machine and Deep Learning Applications

In this chapter, we introduce several applications of machine learning and deep learning in different domains, including sensor and time-series, image and vision, text and natural language processing, relational data, energy, manufacturing, social media, health, security, and Internet-of-Things (IoT) applications. Until 2010, traditional ML models such as SVMs and decision trees have enjoyed successes in various tasks, including handwritten digit classification, face detection, and pattern recognition. Though traditional ML models are easy to interpret, the model's inputs need to be well-designed, handcrafted features. On the other hand, deep learning models circumvent this problem and directly take the raw data as input and provide end-to-end learning capability. There is an unprecedented increase in machine learning and deep learning applications, especially with the emergence of fast mobile devices with access to cloud computing. While cloud computing provides the necessary computational power to train deep learning models, trained models can be easily deployed in the cloud or on embedded devices at the edge of the cloud to carry out the inference.

6.1 SENSOR DATA ANALYTICS

Sensors are everywhere! Sensors enable us to collect a variety of data and also the option to store it in the cloud. However, the primary focus here is to effectively fuse various sensor data and learn patterns from this enormous data. Recent ML and DL algorithms enable us to learn helpful information and knowledge from the raw-sensor data without any human intervention.

A generic sensor signal/data processing framework that includes preprocessing, noise removal, and segmentation is shown in Figure 6.1. The raw sensor signal acquired from sensors like accelerometer or gyro sensor is typically processed in a frame-by-frame or batch mode. This is followed by a preprocessing step where the signal is usually denoised or whitened. In the feature extraction module, important statistical information like mean, variance, and other fundamental characteristics of the signal like zero-crossing and Fourier coefficients are computed. An application of traditional ML follows the feature extraction step. The ML model is trained for a specific task, and later the trained model is used for inference to decide the overall sensor system.

A framework for machine learning-based intrusion detection for wireless sensor networks is provided in [181]. Sensor fault diagnosis is carried out by training SVMs on preprocessed

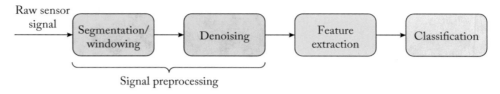

Figure 6.1: A generic signal processing framework which applies preprocessing step prior to feature extraction and classification.

sensor data in [182]. In [94], authors integrate machine learning algorithms on an embedded system for IoT applications. A three-dimensional localization ML algorithm is proposed for wireless sensor nodes in [183]. An in depth-study of modern sensor-fusion techniques to minimize noise and extract efficient features and sensor data analytics using machine learning is provided in [184]. Deep learning also finds its applications in sensor data analytics since they are tailored for end-to-end learning. Blind calibration of sensor networks is performed in [185] using deep learning models. [186] and [187] focus on human activity recognition with multi-channel inertial sensors using DL algorithms.

6.2 MACHINE CONDITION MONITORING

There have been several studies in machine condition monitoring [188, 189] for several applications including electrical motors [190], gear box, bearings [191], production lines, wind turbines [192–194], jet and automobile engines [195], jet wings, and transformers [196]. Machine condition monitoring (MCM) and predictive analysis using ML reduces cost of maintenance and often prevents catastrophic failures. Multiple specialized sensors are often used in MCM including accelerometers, ultrasound, cameras, temperature, chemical, microphones, voltage, and current sensors. Machine learning for turbine MCM was discussed in [194] and vibration condition monitoring in [197]. For rotating machine, gears and bearings Fourier and Wavelet transforms are often used to extract features and ML models are applied for making predictions. Accelerometers are also used for structural monitoring on bridges [198], airplane wings [195], and buildings [199]. In chip manufacturing cameras and vision-based sensors and algorithms as well as precise temperature sensors and control algorithms are used to estimate and/or predict impurities [200]. Several survey papers are available in the literature [201].

6.3 ML IN IMAGE AND VISION

Image and vision applications focus on designing relevant algorithms to capture, identify, and recognize crucial visual information. The field of computer vision (CV) has witnessed a shift in its progress. Earlier techniques in CV were purely based on statistical and rule-based methods. The algorithms consisted of detecting domain-specific feature descriptors (SIFT, SURF, HOG,

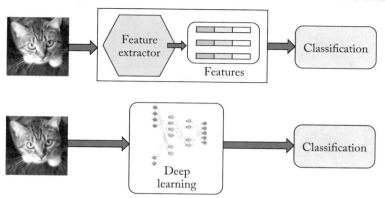

Figure 6.2: Difference between classical computer vision (top figure) and deep learning-based computer vision (bottom figure). Manual feature descriptors are no longer needed.

etc.) for object detection. Feature descriptors like edge detection, corner detection were efficient in identifying informative patches in the images. A bag of features would then be provided as input to classical ML algorithms for training. During inference, the ML algorithm would look for a specific bag of features, and if it exceeded a predefined threshold, then the image would be classified accordingly. Because the CV systems depended on rule-based techniques, they performed sub-optimally. Simple changes in the object shape, scale, orientation, and size would break the aforementioned rule-based systems. Moreover, it is impossible to define such rule-based systems that would incorporate all possible real-world scenarios. Deep learning algorithms entirely replaced classical computer vision techniques since they can extract features.

The efficacy of deep learning models has been demonstrated in many real-life applications, including agriculture, astronomy, biomedicine, transportation, sports, and education. In the food and agricultural sector, DL algorithms are used for food production [202], flood detection, and early disease detection on the leaves of tomato plants [203]. Using historical weather data and meteorological phenomena, fast and accurate weather forecasting [110] is possible. In medicine, it is possible to perform cancer diagnosis [204] and COVID-19 detection [205] from chest x-ray images.

Another major application of deep learning and computer vision is seen in autonomous self-driving cars. Modern DL algorithms are exceptional in object recognition, object localization, and segmentation. Algorithms like YOLO can accurately generate bounding boxes around the object in addition to identifying and recognizing the object [206]. DL models can also be applied to create artistic imagery through image colorization [207], neural style transfer [208], and enhancing image resolutions [209]. Automatic image captioning [210, 211] where a trained ML model will generate a caption that describes the content in the image is achieved through convo-

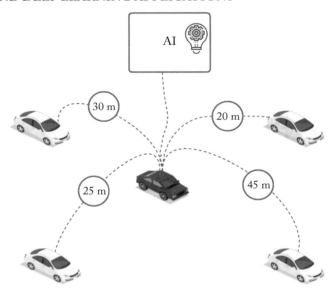

Figure 6.3: Autonomous vehicles integrate a multitude of sensors like LIDAR, RADAR, infrared, and cameras and employ deep learning algorithms to obtain an accurate representation of the surrounding environment.

lutional and recurrent neural networks. Note that the DL model needs to recognize important objects and actions in the image and generate syntactically and semantically correct sentences.

6.4   AUTONOMOUS VEHICLE APPLICATIONS

Autonomous vehicles are in third-phase development and testing and ML/AI has a critical role in many of the navigation and guidance system. There are several applications of ML in autonomous vehicle sensor systems [212]. Scene analysis from multiple camera sensors using ML algorithms has been addressed in [213–215]. In addition, several intelligent radar and lidar applications have been launched [216, 217]. Other challenges include intelligent mobile wireless communication networks [218, 219] and the harnessing of 5G technologies [220]. In addition, AI-enabled localization techniques have been developed for integration in autonomous cars [221]. Finally, imitation learning using deep convolutional neural networks is discussed in [222].

6.5   WIRELESS COMMUNICATIONS ENABLED BY ML

The ability of wireless transceivers to identify automatically channel conditions and apply optimal strategies for channel coding and equalization is critical in emerging standards. In fact,

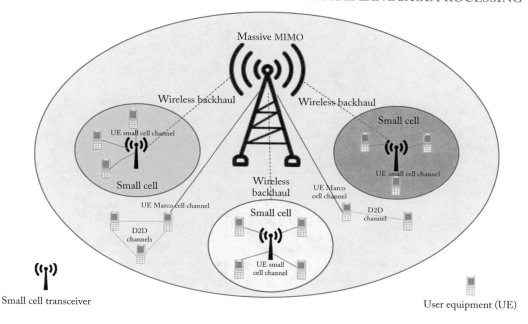

Figure 6.4: Wireless communications applications enabled by machine learning and neural networks.

application of machine learning in communications systems has gained a lot of attention recently. In particular, Massive multiple input multiple output (MIMO) communication systems are critical for increasing the capacity and robustness of emerging wireless networks [223–231]. Emerging wireless networks require intelligent and AI-enabled algorithms for improving data rate, latency, energy efficiency [232], and cost. Improving standards will require millimeter wave (mmWave) advances [233], Heterogeneous Networks [234], and Massive MIMO [235]. Many of the algorithms used in such applications include neural networks and deep learning architectures [236, 237]. In addition, AI-enabled channel coding methods have been proposed in [238]. Application of machine learning in cognitive satellite communications is also an area that is being explored in several research laboratories [239]. Additional work has also been done in the area of radar communications which will be described later in this chapter.

6.6   TEXT AND NATURAL LANGUAGE PROCESSING

Natural Language Processing (NLP) is a crucial part of artificial intelligence. NLP enables machines to understand, interpret, and manipulate human languages. NLP is an interdisciplinary field with ideas and techniques borrowed from linguistics, computational linguistics, speech, and computer science. As witnessed in computer vision, earlier NLP techniques were based on

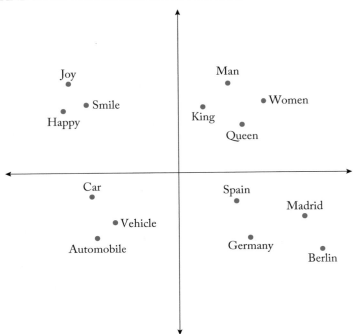

Figure 6.5: The embeddings of a few selected words obtained from the word2vec model [244]. The embedding model preserves the semantic relationship between the words in the latent space.

complex sets of hand-written rules. With the advent of the World Wide Web, a vast amount of natural language data (although unannotated) became available. Now, the field of NLP has also seen some significant breakthroughs because of deep learning. The applications of NLP are everywhere! NLP applications include information retrieval and extraction [240], machine translation [241] (converting from English to French, for example), sentiment analysis [242], spam filtering [7], auto-predicting the next word [243] in a sentence, etc.

Basic NLP tasks include tokenization of words, parsing, and stemming. Popular and recent deep learning techniques include word embedding models that capture the semantic properties of words. For example, the Word2Vec [244, 245] model embeds each distinct word as a vector in latent space. The low-dimensional vectors, also known as *embeddings*, are obtained after training the model. Similar words have vectors grouped closed together in the latent space. For example, words like car, automobile, and vehicle might be clustered together in one corner, while angry, happy, and sad might be clustered together in another corner. Word2Vec also preserves the relative meaning. Not only will Madrid, London, Paris, and Berlin cluster near each other, but they will each have similar distances in vector space to the countries whose capitals are, i.e., Spain - Madrid = Rome - Italy. The embeddings for a few word vectors are shown in

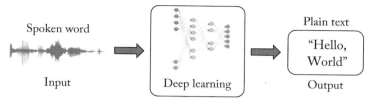

Figure 6.6: The automatic speech recognition engine takes in spoken work as input and outputs a plain text.

Figure 6.5. Transformer architecture with dot-product attention mechanism [246] replaced the standard recurrent neural network architectures for NLP applications. Very recently, OpenAI released a generative pre-training (GPT-3) model that achieves state-of-the-art results for many NLP tasks.

6.7   SPEECH AND AUDIO DATA

Many of the basic ML algorithms were initially applied in the field of speech coding, namely in vector quantization [247, 248]. The K-means algorithm was used to design codebooks for the code excited linear predictor (CELP) [249] that essentially enabled digital cellphone and low bit rate military communications [250]. ML was also applied in sinusoidal coding of speech [251]. Additional compression systems that employ machine learning have been reported in [252–254]. These systems have used machine learning to extend the bandwidth of narrow-band compression algorithms. *Phoneme* is the basic unit of speech, sequences of many such phonemes form a spoken word. Hidden Markov Models (HMMs) [255] were used earlier for speech recognition systems. HMMs are statistical models that output a sequence of symbols (phonemes in the case of speech). HMMs are specifically used in speech recognition [256] because a human speech signal is considered quasi-stationary (short-time stationary signal). Hence, speech is thought of as being generated by a Markovian model for stochastic purposes. HMMs are trained using cepstral coefficients obtained by computing Fourier transform for a short-time window of the speech signal (20 milliseconds, for example).

One of the most popular application areas of deep learning concerning speech is voice search and voice-activated intelligent assistants. Personal digital voice assistants like Siri and Ok Google on our mobile phones can recognize speech with almost 100% accuracy. Generating realistic human-sounding speech or other audio data is now feasible through DeepMind's WaveNet. WaveNet [257] is a deep neural network that employs dilated convolutional layers to produce conditioned audio samples as its output. Other applications in speech include automatic speech recognition (Figure 6.6), text-to-speech synthesis, artificial speech production, speech diarization [258, 259], and emotion recognition [260, 261]. There have also been several ML applications in music recognition [262], query-by-humming [263], and environmental sound detection [264].

6.8 GRAPH AND RELATIONAL DATA

Graph data structures are ideal for modeling a complex system of relations and interactions. Many real-world data are naturally in the form of graphs, including transportation networks, biological networks, knowledge graphs, etc. Thanks to graph-based machine learning, the semi-supervised learning paradigm has seen tremendous progress. For example, algorithms like PageRank (Google search engine) rank web pages according to their importance for a given query. Traditional graph-based machine learning techniques were used to perform node prediction, link prediction, collaborative filtering, and identifying influential nodes. Node embeddings, the key to solving the aforementioned tasks, are obtained through optimization algorithms or matrix factorization methods.

Recently, proposed graph neural network (GNN) formalisms have seen unprecedented success in scientific domains like chemistry, drug design, and fundamental science. A GNN was trained to predict interactions between proteins from their 3D structure [265]. AlphaFold solved a challenging problem in bioinformatics where a $3D$ protein structure is predicted from the amino acid sequence [266]. In [267], a GNN-based pipeline is used to discover new antibiotic drugs.

6.9 TINY AND EMBEDDED MACHINE LEARNING

Tiny Machine Learning (TinyML) [268, 269] is the intersection of ML algorithms on embedded system platforms, edge processing, and IoT devices. TinyML is also referred to as *machine learning on edge* since the algorithmic processing occurs at the device level or local level. TinyML is rapidly becoming more accessible. The focus of TinyML is on implementing ML algorithms in ultra-low power systems. Traditionally, smartphones and devices on the edge transferred generated data to a centralized cloud system. The ML algorithms hosted on a cloud, when trained on the data, would provide valuable insights, and the learned knowledge would then be transmitted back to the embedded systems. However, with TinyML, it is now possible to deploy and train ML algorithms on IoT devices and generate inference without ever sending the data to the cloud [270]. Algorithms embedded on an MCU running an RTOS can guarantee real-time performance. This is advantageous for several reasons, such as data privacy, reduced transmission cost, and improved power and latency. In addition, this allows for a decentralized system infrastructure rather than a centralized cloud system.

TinyML creates the potential for numerous real-life applications. For example, smartphones react to wake words such as "Ok Google." Recently, a Google team showed that an ML model less than 14 kilobytes in size could detect wake words. TinyML-based smart sensors are being employed for creating a sustainable food future [271]. TinyML-based models are used to

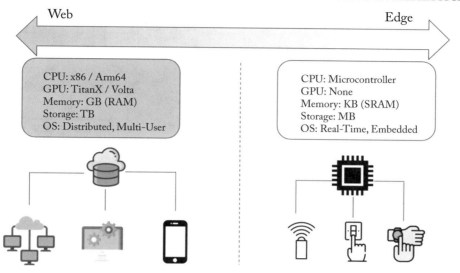

Figure 6.7: Technological differences in cloud-based computing and edge computing.

increase crop yields by continuously monitoring important attributes like pH level, humidity, and carbon dioxide levels. Though TinyML is very promising, there are significant challenges that need to be overcome. Embedded devices have limited memory for processing and storage. It is also difficult to debug and troubleshoot hundreds of TinyML devices, unlike a centralized ML model in the cloud. Other technological challenges are summarized in Figure 6.7. A forum for TinyML work particularly involving industry contributions in edge ML is given in [272, 273].

6.10 MACHINE LEARNING IN HEALTHCARE

Machine learning algorithms are powerful pattern recognition engines and can sift through a large number of patient data records. CNN-based deep learning algorithm trained on retinal images can identify diabetic retinopathy [274]. Similarly, other deep learning models can diagnose skin cancer [275] and cancerous tumors from mammograms. In addition to using ML algorithms in predictive analytics, they are also used to suggest alternative treatments and therapies. Several companies employ ML in their drug discovery pipeline to design new drug-like molecules with desired pharmaceutical properties. LSTM-based neural network models are employed to generate new drug-analogs of varying molecular weight [276]. Omics data such as proteomics, genomics, metabolomics, etc., are integrated to develop a more holistic model of the inner workings of biological systems using ML, enabling the discovery of new biomarkers [277]. Ligand biological activity is predicted by utilizing a set of active and inactive data by discovering significant chemical differences [278]. Several innovations also have been reported in the area of ML for diagnosis of abnormal speech, dysarthria, and diagnosis of neurological

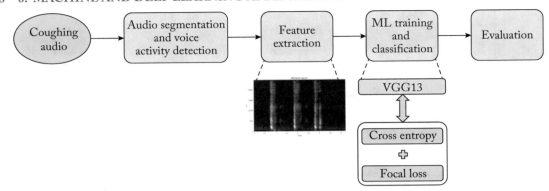

Figure 6.8: An overview of the system used for COVID-19 detection using cough samples. The system uses a unique combination of cross entropy and focal loss functions which yield improved performance with COVID-19 audio.

diseases [279–281]. More recently, there have been applications of ML for COVID-19 cough detection and classification [282, 283], as shown in Figure 6.8.

6.11   MACHINE LEARNING IN ENERGY APPLICATIONS

There have been several studies in energy forecasting [142], solar fault detection, photovoltaic (PV), topology reconnection [284], and solar panel shading prediction [285]. In fault detection, the research included the application of k-means [286] on voltage-current data obtained from monitoring devices. These studies later began using neural networks to detect and classify PV faults using NREL data [287], as shown in Figure 6.9. At the same time, there have been several neural network studies [288] on topology reconfiguration, which have shown that they can elevate the power output by more than 10% under shading conditions. Research on using ML and vision algorithms has also produced promising results for shading prediction in utility-scale solar plants. A summary of many such methodologies is given in [289]. Recent techniques that use pruned neural networks for PV fault detection have been reported in [290] and simulations of fault detection using quantum machine learning have been reported in [291]. Additional recent work in this area can be found in [292] and [293].

6.12   MACHINE LEARNING IN DEFENSE AND SECURITY APPLICATIONS

There exist several critical military applications that employ machine learning, and artificial intelligence [294]. Military imaging systems deployed for threat monitoring and situational awareness can detect, identify, and can also track objects or entities of interest [295]. Location of underwater mines is possible by using a deep convolutional neural network with an accuracy

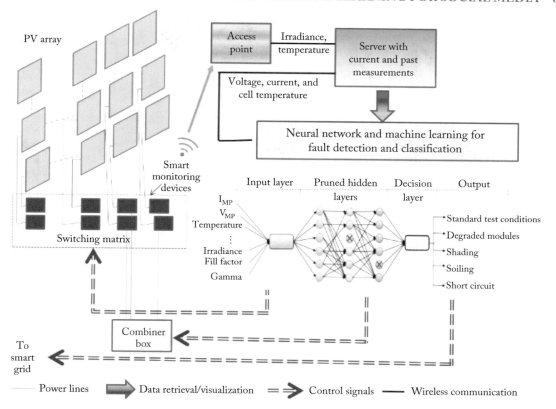

Figure 6.9: Smart solar array monitoring system with shading prediction, topology optimization, and fault diagnosis systems.

of 90% [296, 297]. Another prime example of AI in the military can be seen in cybersecurity. AI-enabled systems can proactively detect malicious activity and stop attacks [298]. Finally, intelligent radar applications have also been reported in [299, 300].

6.13   MACHINE LEARNING FOR SOCIAL MEDIA

Much of the digital data generated every day is from Social Media Platforms like Facebook, Twitter, Instagram, and Youtube. Apparently, in 2020, people created 1.7 MB of data every second [301]. People create more than 95 millions posts on Instagram per day and 32 million Facebook posts per minute and close to 6000 tweets per second. Scalable deep learning models can quickly swift through the enormous data and can gain better insights, as shown in Figure 6.10. By leveraging the knowledge acquired from the DL models, it is possible to accurately market products and place relevant advertisements [302]. Marketing companies can analyze the

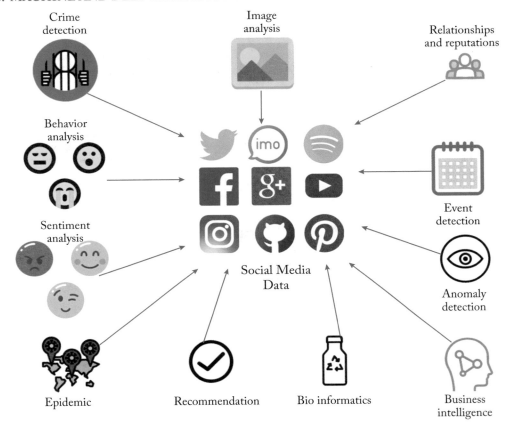

Figure 6.10: Social media analytics, applications, and benefits using data-driven machine learning and deep learning algorithms.

comments from user to recognize user sentiments [303] for the products [304]. Many data-driven algorithms can track the trends and rise in popularity of the products by tracking the comments, posts and even images. AI platforms can even identify user accounts which are popular and influential and choosing them for product promotions [305].

6.14 MACHINE LEARNING IN ENTERTAINMENT

Many content creators and media enterprises make use of artificial intelligence for enhancing user experience [306] and personalization [307, 308]. Companies like Netflix with more than 200 million customers integrate advanced data analytics in their pipelines for user recommendations. Netflix collects every possible information such as the date and time a user watched a show, the device used, how many times the show was paused, and how long it takes the user to finish a show [309]. They make use of ML models to even generate the thumbnails [310] using thousands of frames of the video. It identifies which of the thumbnails are more likely to be clicked and watched.

6.15   MACHINE LEARNING IN MANUFACTURING

The cost of fixing an equipment failure is often high and time consuming. Manufacturers may have to spend too much time fixing breakdowns instead of allocating resources for planned maintenance and production. However, fortunately, manufacturing industry today is experiencing a never seen increase in available data [311] comprising of different formats like sensor data,and environmental data, and equipment operating parameters. Several supervised and unsupervised machine learning techniques can make use of the data to perform predictive maintenance by identifying equipment failures even before they occur [312]. It is also shown that combining multiple models instead of single model can significantly improve prediction accuracies [313]. Apart from predictive maintenance, AI tools are also employed for improving predictive quality and yield. ML algorithms identify root cause of the process-driven production losses. Some manufactures are using AI platforms to improve demand forecasting accuracies to minimize energy costs and predict negative price variances [314].

6.16   QUANTUM MACHINE LEARNING

Quantum machine learning (QML) is currently a "hot" area of interest because of the promise of enormous computation speeds in quantum computing [315]. QML algorithm cover supervised and unsupervised methods for classification and clustering [316]. Unsupervised QML has the potential for improved speed by performing tasks simultaneously and leveraging quantum superposition. For example, in K-mean clustering applications, the distances to all the centroids can be computed simultaneously. Some core QML algorithms appeared in the literature as follows: K-nearest neighbor [317], K-means clustering, and decision tree. Support vector machines with quantum computing were reported in [318]. There have been several applications documented in this field [315] including, energy, voice and image processing, communications, intelligent networks, radar, and encryption. In solar array fault detection, quantum neural network simulations were developed (see Figure 6.11) and a two-qubit ANN hybrid implementation was examined in [291]. This study reported simulations highlighting trade-offs of qubit precision versus quantum noise. We note that access to quantum computing facilities is currently limited and expensive, however, there are several quantum simulators [316–318] that have been developed and are available for research and education.

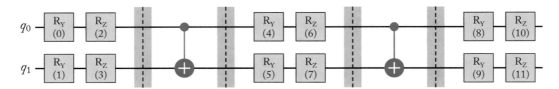

Figure 6.11: Sample simulation of a quantum neural network used for solar array fault detection [291].

CHAPTER 7

Conclusion and Future Directions

An introduction to machine learning principles with a comprehensive literature review was presented in this book. The book is written to provide knowledge and bibliography on machine learning and neural networks concepts to a reader with minimal background in machine learning. We started with the fundamental learning paradigms in ML and explored the sub-categories in each. Supervised learning, unsupervised learning, and semi-supervised learning are the three main categories in machine learning. Most of the existing algorithms can be classified into one of the above three categories. The book is organized to cover algorithms and concepts first. It later describes the applications of ML algorithms in various fields, including signal processing, image and computer vision, natural language processing, speech and audio processing, energy, health, security, and defense applications.

Chapter 1 started with an introduction to ML. It also presented a brief history of the progress seen in the field of ML. The rise in the popularity of ML and DL can be attributed to the availability of big data, computing resources, and software frameworks. The hierarchical relationship among AI, ML, and DL was also explored in this chapter.

Chapter 2 focused only on supervised ML. Supervised learning has access to ground truth or true labels in addition to the data. A supervised learning algorithm aims to learn a mapping function from input to the ground truth. Different regression and classification algorithms in supervised ML were presented in this chapter and several references were cited.

Chapter 3 discussed the importance of unsupervised learning in the absence of true labels to the data. The common problems encountered in unsupervised learning are cluster analysis and dimensionality reduction. The chapter introduced nonparametric and parametric models often employed in unsupervised learning. The commonly used clustering algorithms such as the K-means algorithm, spectral clustering, and Gaussian Mixture Models were discussed. The last section of the chapter focused on dimensionality reduction algorithms in machine learning. In the chapter we cited several algorithms and structures for further reading.

Chapter 4 focused on semi-supervised learning, a paradigm where a lot of unlabeled data is used with a limited amount of labeled data. It explains how a large amount of unlabeled data can be used, even with a few labeled ones. Most of the chapter is focused on graph-based machine learning and showcasing different approaches in the literature. Positive unlabeled learning, a sub-paradigm of semi-supervised learning, is introduced at the end.

Chapter 5 is a brief introduction to deep learning algorithms. It starts with the basics of neural networks and their training through a back-propagation algorithm. Neural networks as such may be prone to overfitting, and many approaches exist in practice to avoid or reduce overfitting. The convolutional neural network architecture works exceptionally well with spatial data such as images, whereas recurrent neural networks work well on temporal or time-series data. Auto-encoders and generative adversarial networks are applications of neural networks for unsupervised representation learning.

Chapter 6 is dedicated to current ML and DL applications. We introduced several machine learning applications and deep learning in different data domains, including sensor and time-series data, image and vision, text and natural language processing, energy, defense, machine condition monitoring, sensor calibration, relational data, and Internet-of-Things (IoT) applications. Our coverage of applications is extensive but not exhaustive. There are several applications emerging as new ML tools and architectures are developed, and in particular there are quite a few new paradigms in the embedded and TinyML space. In addition, the emergence of quantum machine learning was examined. QML is a technology that promises to vastly accelerate solutions for big data analytics.

7.1   FURTHER READING

In this section, we recommend denser books for further reading. The recommendations provided are for mathematics for ML, traditional ML theory, statistical ML, DL books, and ML for practitioners.

- **Mathematics for ML**: For those readers who want to gain an in-depth understanding of the fundamental concepts and theory behind ML are recommended to go through *Mathematics for Machine Learning* book [319]. This is an open-source and freely available book that provides an in-depth understanding of linear algebra, vector calculus, probability theory, and continuous optimization. Apart from the math concepts, the book also provides examples of machine learning algorithms that use the mathematical foundations.

- **ML Theory**: Graduate students and other readers who want to learn ML from its theoretical principles are encouraged to read *Understanding Machine Learning—from Theory to Algorithms* [320]. Necessary proofs on many theorems in ML are provided throughout the book.

- **Statistical ML**: We recommend *The Elements of Statistical Learning* [3] for the readers with a limited background in statistics. The book provides a statistical learning approach to ML algorithms and starts with a statistical decision theory.

- **Deep Learning**: To learn advanced concepts in deep learning and artificial neural networks, readers are recommended to go through *Deep Learning* book [10]. This book also covers DL applications in computer vision and many other fields.

- **ML for Practitioners**: For software engineers and practitioners who favor hands on approach to ML are advised to go through *Introduction to ML with Python* [321]. We also recommend online documentation [322] for ML libraries such as *Scikit-Learn*. Practitioners are also encouraged to learn deep learning frameworks such as Pytorch [323], Keras [324], and TensorFlow [322].

Bibliography

[1] Uday Shankar Shanthamallu, Andreas Spanias, Cihan Tepedelenlioglu, and Mike Stanley. A brief survey of machine learning methods and their sensor and IoT applications. *8th International Conference on Information, Intelligence, Systems and Applications (IISA)*, pages 1–8, IEEE, 2017. DOI: 10.1109/iisa.2017.8316459 1

[2] Christian Robert. *Machine Learning, a Probabilistic Perspective*. The MIT Press, 2014. DOI: 10.1080/09332480.2014.914768 1

[3] Jerome Friedman, Trevor Hastie, Robert Tibshirani, et al. *The Elements of Statistical Learning*, vol. 1. Springer Series in Statistics, New York, 2001. DOI: 10.1007/978-0-387-21606-5 1, 74

[4] Christopher M. Bishop. *Pattern Recognition and Machine Learning*. Springer, 2006. 1

[5] Yann LeCun, Bernhard Boser, John S. Denker, Donnie Henderson, Richard E. Howard, Wayne Hubbard, and Lawrence D. Jackel. Backpropagation applied to handwritten zip code recognition. *Neural Computation*, 1(4):541–551, 1989. DOI: 10.1162/neco.1989.1.4.541 1, 50, 52

[6] Dengsheng Lu and Qihao Weng. A survey of image classification methods and techniques for improving classification performance. *International Journal of Remote Sensing*, 28(5):823–870, 2007. DOI: 10.1080/01431160600746456 1

[7] Emmanuel Gbenga Dada, Joseph Stephen Bassi, Haruna Chiroma, Adebayo Olusola Adetunmbi, Opeyemi Emmanuel Ajibuwa, et al. Machine learning for email spam filtering: Review, approaches, and open research problems. *Heliyon*, 5(6):e01802, 2019. DOI: 10.1016/j.heliyon.2019.e01802 1, 64

[8] Carson Kai-Sang Leung, Richard Kyle MacKinnon, and Yang Wang. A machine learning approach for stock price prediction. *Proc. of the 18th International Database Engineering and Applications Symposium*, pages 274–277, 2014. DOI: 10.1145/2628194.2628211 1

[9] Yann LeCun, Yoshua Bengio, and Geoffrey Hinton. Deep learning. *Nature*, 521(7553):436–444, 2015. DOI: 10.1038/nature14539 1, 43

[10] Ian Goodfellow, Yoshua Bengio, Aaron Courville, and Yoshua Bengio. *Deep Learning*, vol. 1. MIT Press Cambridge, 2016. DOI: 10.1038/nature14539 1, 75

[11] Halbert White. Learning in artificial neural networks: A statistical perspective. *Neural Computation*, 1(4):425–464, 1989. DOI: 10.1162/neco.1989.1.4.425 1

[12] Ian Rogers. The Google Pagerank Algorithm and How it Works, 2002. 1

[13] Greg Linden, Brent Smith, and Jeremy York. Amazon.com recommendations: Item-to-item collaborative filtering. *IEEE Internet Computing*, 7(1):76–80, 2003. DOI: 10.1109/mic.2003.1167344 1

[14] Robert M. Bell and Yehuda Koren. Lessons from the Netflix prize challenge. *ACM Sigkdd Explorations Newsletter*, 9(2):75–79, 2007. DOI: 10.1145/1345448.1345465 1

[15] Thomas Bayes. LII. An essay towards solving a problem in the doctrine of chances. By the late Rev. Mr. Bayes, F. R. S. communicated by Mr. Price, in a letter to John Canton, A. M. F. R. S. *Philosophical Transactions of the Royal Society of London*, (53):370–418, 1763. DOI: 10.1098/rstl.1763.0053 2

[16] Alan M. Turing. Computing machinery and intelligence. *Parsing the Turing Test*, pages 23–65, Springer, 2009. 2

[17] Frank Rosenblatt. The perceptron: A probabilistic model for information storage and organization in the brain. *Psychological Review*, 65(6):386, 1958. DOI: 10.1037/h0042519 2, 43

[18] Marvin L. Minsky and Seymour A. Papert. *Perceptrons: Expanded edition*, 1988. 2

[19] Robert Gray. Vector quantization. *IEEE ASSP Magazine*, 1(2):4–29, 1984. DOI: 10.1109/MASSP.1984.1162229 2

[20] Diederik P. Kingma and Max Welling. An introduction to variational autoencoders. *ArXiv Preprint ArXiv:1906.02691*, 2019. DOI: 10.1561/2200000056 2, 55

[21] Ian J. Goodfellow, Jean Pouget-Abadie, Mehdi Mirza, Bing Xu, David Warde-Farley, Sherjil Ozair, Aaron Courville, and Yoshua Bengio. Generative adversarial networks, 2014. DOI: 10.1145/3422622 2, 55

[22] Norman P. Jouppi, Cliff Young, Nishant Patil, David Patterson, Gaurav Agrawal, Raminder Bajwa, Sarah Bates, Suresh Bhatia, Nan Boden, Al Borchers, et al. In-datacenter performance analysis of a tensor processing unit. In *Proc. of the 44th Annual International Symposium on Computer Architecture*, pages 1–12, 2017. DOI: 10.1145/3079856.3080246 2

[23] Jaime G. Carbonell, Ryszard S. Michalski, and Tom M. Mitchell. An overview of machine learning. *Machine Learning*, pages 3–23, 1983. DOI: 10.1007/978-3-662-12405-5_1 4

[24] Mu Li, Tong Zhang, Yuqiang Chen, and Alexander J. Smola. Efficient mini-batch training for stochastic optimization. In *Proc. of the 20th ACM SIGKDD International Conference on Knowledge Discovery and Data Mining*, pages 661–670, 2014. DOI: 10.1145/2623330.2623612 4

[25] Ning Qian. On the momentum term in gradient descent learning algorithms. *Neural Networks*, 12(1):145–151, 1999. DOI: 10.1016/s0893-6080(98)00116-6 4

[26] Aleksandar Botev, Guy Lever, and David Barber. Nesterov's accelerated gradient and momentum as approximations to regularised update descent. *International Joint Conference on Neural Networks (IJCNN)*, pages 1899–1903, IEEE, 2017. DOI: 10.1109/ijcnn.2017.7966082 4

[27] Richard S. Sutton, Andrew G. Barto, and O. Barana. Book reviews-reinforcement learning: An introduction. *IEEE Transactions on Neural Networks*, 16(1):285–285, 2005. DOI: 10.1109/TNN.1998.712192 5

[28] Csaba Szepesvári. Algorithms for reinforcement learning. *Synthesis Lectures on Artificial Intelligence and Machine Learning*, 4(1):1–103, 2010. DOI: 10.2200/s00268ed1v01y201005aim009 5

[29] Sudharsan Ravichandiran. *Hands-on Reinforcement Learning with Python: Master Reinforcement and Deep Reinforcement Learning Using OpenAI Gym and TensorFlow*. Packt Publishing Ltd., 2018. 5

[30] Md Afzal Hossan, Sheeraz Memon, and Mark A. Gregory. A novel approach for MFCC feature extraction. *4th International Conference on Signal Processing and Communication Systems*, pages 1–5, IEEE, 2010. DOI: 10.1109/icspcs.2010.5709752 5

[31] Yuyao Wang, Zhengming Li, Long Wang, Min Wang, et al. A scale invariant feature transform based method. *Journal of Information Hiding and Multimedia Signal Processing*, 4(2):73–89, 2013. 5

[32] William T. Freeman and Michal Roth. Orientation histograms for hand gesture recognition. *International Workshop on Automatic Face and Gesture Recognition*, 12:296–301, IEEE Computer Society, Washington, DC, 1995. 5

[33] R. Mitchell, J. Michalski, and T. Carbonell. *An Artificial Intelligence Approach*. Springer, 2013. DOI: 10.1007/978-3-662-12405-5 6

[34] Pádraig Cunningham, Matthieu Cord, and Sarah Jane Delany. Supervised learning. *Machine Learning Techniques for Multimedia*, pages 21–49, Springer, 2008. DOI: 10.1007/978-3-540-75171-7_2 9

[35] Aristomenis S. Lampropoulos and George A. Tsihrintzis. Machine learning paradigms. *Applications in Recommender Systems*. Springer International Publishing, Switzerland, 2015. DOI: 10.1007/978-3-319-19135-5 10

[36] Josep Lluis Berral-García. A quick view on current techniques and machine learning algorithms for big data analytics. *18th International Conference on Transparent Optical Networks (ICTON)*, pages 1–4, IEEE, 2016. DOI: 10.1109/ICTON.2016.7550517 10

[37] Norman Matloff. *Statistical Regression and Classification: From Linear Models to Machine Learning*. CRC Press, 2017. DOI: 10.1201/9781315119588 10, 11

[38] Sotiris B. Kotsiantis, I. Zaharakis, and P. Pintelas. Supervised machine learning: A review of classification techniques. *Emerging Artificial Intelligence Applications in Computer Engineering*, 160(1):3–24, 2007. 10

[39] David M. Allen. Mean square error of prediction as a criterion for selecting variables. *Technometrics*, 13(3):469–475, 1971. DOI: 10.1080/00401706.1971.10488811 11

[40] Cort J. Willmott and Kenji Matsuura. Advantages of the mean absolute error (MAE) over the root mean square error (RMSE) in assessing average model performance. *Climate Research*, 30(1):79–82, 2005. DOI: 10.3354/cr030079 11

[41] Douglas C. Montgomery, Elizabeth A. Peck, and G. Geoffrey Vining. *Introduction to Linear Regression Analysis*, vol. 821. John Wiley & Sons, 2012. 11

[42] David A. Freedman. *Statistical Models: Theory and Practice*. Cambridge University Press, 2009. DOI: 10.1017/cbo9780511815867 11

[43] Alvin C. Rencher. *Methods of Multivariate Analysis*, vol. 492. John Wiley & Sons, 2003. DOI: 10.1002/0471271357 11

[44] Douglas M. Bates and Donald G. Watts. *Nonlinear Regression Analysis and its Applications*, vol. 2. Wiley New York, 1988. DOI: 10.1002/9780470316757 12

[45] Eva Ostertagová. Modelling using polynomial regression. *Procedia Engineering*, 48:500–506, 2012. DOI: 10.1016/j.proeng.2012.09.545 12

[46] Raymond E. Wright. Logistic regression. In L. G. Grimm and P. R. Yarnold (Eds.), *Reading and Understanding Multivariate Statistics, American Psychological Association*, pages 217–244, 1995. 13

[47] Juliana Tolles and William J. Meurer. Logistic regression: Relating patient characteristics to outcomes. *Jama*, 316(5):533–534, 2016. DOI: 10.1001/jama.2016.7653 13

[48] Corinna Cortes and Vladimir Vapnik. Support-vector networks. *Machine Learning*, 20(3):273–297, 1995. DOI: 10.1007/bf00994018 15

[49] Harris Drucker, Chris J. C. Burges, Linda Kaufman, Alex Smola, Vladimir Vapnik, et al. Support vector regression machines. *Advances in Neural Information Processing Systems*, 9:155–161, 1997. 15

[50] Asa Ben-Hur, David Horn, Hava T. Siegelmann, and Vladimir Vapnik. Support vector clustering. *Journal of Machine Learning Research*, 2(Dec):125–137, 2001. DOI: 10.4249/scholarpedia.5187 15

[51] Christopher J. C. Burges. A tutorial on support vector machines for pattern recognition. *Data Mining and Knowledge Discovery*, 2(2):121–167, 1998. DOI: 10.1023/A:1009715923555 17

[52] Bernhard E. Boser, Isabelle M. Guyon, and Vladimir N. Vapnik. A training algorithm for optimal margin classifiers. In David Haussler, Ed., *Proc. of the 5th Annual Workshop on Computational Learning Theory (COLT'92)*, pages 144–152, ACM Press, Pittsburgh, PA, July 1992. DOI: 10.1145/130385.130401 17

[53] Kemal Polat and Salih Güneş. Breast cancer diagnosis using least square support vector machine. *Digital Signal Processing*, 17(4):694–701, 2007. DOI: 10.1016/j.dsp.2006.10.008 17

[54] Yunqiang Chen, Xiang Sean Zhou, and Thomas S. Huang. One-class SVM for learning in image retrieval. *Proc. International Conference on Image Processing (Cat. no. 01CH37205)*, 1:34–37, IEEE, 2001. DOI: 10.1109/icip.2001.958946 17

[55] Kun-Lun Li, Hou-Kuan Huang, Sheng-Feng Tian, and Wei Xu. Improving one-class SVM for anomaly detection. *Proc. of the International Conference on Machine Learning and Cybernetics (IEEE Cat. no.03EX693)*, 5:3077–3081, 2003. DOI: 10.1109/icmlc.2003.1260106 17

[56] Gautam M. Borkar, Leena H. Patil, Dilip Dalgade, and Ankush Hutke. A novel clustering approach and adaptive SVM classifier for intrusion detection in WSN: A data mining concept. *Sustainable Computing: Informatics and Systems*, 23:120–135, 2019. DOI: 10.1016/j.suscom.2019.06.002 17

[57] Evelyn Fix and J. L. Hodges. Discriminatory analysis. Nonparametric discrimination: Consistency properties. *International Statistical Review/Revue Internationale de Statistique*, 57(3):238–247, 1989. DOI: 10.2307/1403797 18

[58] Naomi S. Altman. An introduction to kernel and nearest-neighbor nonparametric regression. *The American Statistician*, 46(3):175–185, 1992. DOI: 10.1080/00031305.1992.10475879 18

[59] Warren B. Powell. *Approximate Dynamic Programming: Solving the Curses of Dimensionality*, vol. 703. John Wiley & Sons, 2007. DOI: 10.1002/9780470182963 19

[60] Yunsheng Song, Jiye Liang, Jing Lu, and Xingwang Zhao. An efficient instance selection algorithm for k nearest neighbor regression. *Neurocomputing*, 251:26–34, 2017. DOI: 10.1016/j.neucom.2017.04.018 19

[61] Irina Rish et al. An empirical study of the naive Bayes classifier. *IJCAI Workshop on Empirical Methods in Artificial Intelligence*, 3:41–46, 2001. 19

[62] David J. Hand and Keming Yu. Idiot's Bayes—not so stupid after all? *International Statistical Review*, 69(3):385–398, 2001. DOI: 10.1111/j.1751-5823.2001.tb00465.x 19

[63] Ali Haghpanah Jahromi and Mohammad Taheri. A non-parametric mixture of Gaussian naive Bayes classifiers based on local independent features. *Artificial Intelligence and Signal Processing Conference (AISP)*, pages 209–212, IEEE, 2017. DOI: 10.1109/aisp.2017.8324083 20

[64] Ashraf M. Kibriya, Eibe Frank, Bernhard Pfahringer, and Geoffrey Holmes. Multinomial naive Bayes for text categorization revisited. *Australasian Joint Conference on Artificial Intelligence*, pages 488–499, Springer, 2004. DOI: 10.1007/978-3-540-30549-1_43 20

[65] Lior Rokach and Oded Maimon. Decision trees. *Data Mining and Knowledge Discovery Handbook*, pages 165–192, Springer, 2005. DOI: 10.1007/0-387-25465-x_9 20

[66] Carl Kingsford and Steven L. Salzberg. What are decision trees? *Nature Biotechnology*, 26(9):1011–1013, 2008. DOI: 10.1038/nbt0908-1011 20

[67] Xindong Wu, Vipin Kumar, J. Ross Quinlan, Joydeep Ghosh, Qiang Yang, Hiroshi Motoda, Geoffrey J. McLachlan, Angus Ng, Bing Liu, S. Yu Philip, et al. Top 10 algorithms in data mining. *Knowledge and Information Systems*, 14(1):1–37, 2008. DOI: 10.1007/s10115-007-0114-2 20

[68] John Mingers. An empirical comparison of pruning methods for decision tree induction. *Machine Learning*, 4(2):227–243, 1989. DOI: 10.1023/A:1022604100933 20

[69] Horace B. Barlow. Unsupervised learning. *Neural Computation*, 1(3):295–311, 1989. DOI: 10.1162/neco.1989.1.3.295 23

[70] Zoubin Ghahramani. Unsupervised learning. *Summer School on Machine Learning*, pages 72–112, Springer, 2003. DOI: 10.1007/978-3-540-28650-9_5 23

[71] Trevor Hastie, Robert Tibshirani, and Jerome Friedman. Unsupervised learning. *The Elements of Statistical Learning*, pages 485–585, Springer, 2009. DOI: 10.1007/978-0-387-84858-7_14 23

[72] Charu C. Aggarwal and C. K. Reddy. An introduction to cluster analysis, 2013. DOI: 10.1007/978-1-4614-6396-2_1 23

[73] Alboukadel Kassambara. *Practical Guide to Cluster Analysis in R: Unsupervised Machine Learning*, vol. 1. Sthda, 2017. 23

[74] Laurens Van Der Maaten, Eric Postma, and Jaap Van den Herik. Dimensionality reduction: A comparative. *Journal of Machine Learning and Research*, 10(66–71):13, 2009. 23

[75] John A. Lee and Michel Verleysen. *Nonlinear Dimensionality Reduction*. Springer Science & Business Media, 2007. DOI: 10.1007/978-0-387-39351-3 23, 30

[76] Vilijandas Bagdonavičius, Julius Kruopis, and Mikhail Stepanovich Nikulin. *Non-Parametric Tests for Complete Data*. Wiley Online Library, 2011. DOI: 10.1002/9781118557716 23

[77] Gregory W. Corder and Dale I. Foreman. *Nonparametric Statistics: A Step-by-Step Approach*. John Wiley & Sons, 2014. DOI: 10.1002/9781118165881 23

[78] Friedrich Liese and Klaus-J. Miescke. Statistical decision theory: Estimation. *Testing, and Selection*, 2008. DOI: 10.1007/978-0-387-73194-0_3 24

[79] Peter J. Bickel and Kjell A. Doksum. *Mathematical Statistics: Basic Ideas and Selected Topics, Volumes I–II Package*. CRC Press, 2015. DOI: 10.1201/9781315369266 24

[80] Aristidis Likas, Nikos Vlassis, and Jakob J. Verbeek. The global k-means clustering algorithm. *Pattern Recognition*, 36(2):451–461, 2003. DOI: 10.1016/s0031-3203(02)00060-2 24

[81] John A. Hartigan and Manchek A. Wong. A k-means clustering algorithm. *Journal of the Royal Statistical Society: Series C (Applied Statistics)*, 28(1):100–108, 1979. DOI: 10.2307/2346830 24

[82] Chris Piech. The k-means algorithm.

[83] Purnima Bholowalia and Arvind Kumar. Ebk-means: A clustering technique based on elbow method and k-means in WSN. *International Journal of Computer Applications*, 105(9):2014. DOI: 10.5120/18405-9674 25

[84] Junjie Wu. *Advances in K-means Clustering: A Data Mining Thinking*. Springer Science & Business Media, 2012. DOI: 10.1007/978-3-642-29807-3 25

[85] K. Krishna and M. Narasimha Murty. Genetic k-means algorithm. *IEEE Transactions on Systems, Man, and Cybernetics, Part B (Cybernetics)*, 29(3):433–439, 1999. DOI: 10.1109/3477.764879 25

[86] Ulrike Von Luxburg. A tutorial on spectral clustering. *Statistics and Computing*, 17(4):395–416, 2007. DOI: 10.1007/s11222-007-9033-z 26

[87] Andrew Y. Ng, Michael I. Jordan, Yair Weiss, et al. On spectral clustering: Analysis and an algorithm. *Advances in Neural Information Processing Systems*, 2:849–856, 2002. 26, 37

[88] Fan R. K. Chung and Fan Chung Graham. *Spectral Graph Theory*. Number 92. American Mathematical Society, 1997. DOI: 10.1090/cbms/092 26

[89] Raymond Hon-Fu Chan and Xiao-Qing Jin. *An Introduction to Iterative Toeplitz Solvers*. SIAM, 2007. DOI: 10.1137/1.9780898718850 26

[90] Douglas A. Reynolds. Gaussian mixture models. *Encyclopedia of Biometrics*, 741:659–663, 2009. DOI: 10.1007/978-0-387-73003-5_196 27

[91] Noam Shental, Aharon Bar-Hillel, Tomer Hertz, and Daphna Weinshall. Computing Gaussian mixture models with em using equivalence constraints. *Advances in Neural Information Processing Systems*, 16(8):465–472, 2004. 27

[92] Nathaniel R. Goodman. Statistical analysis based on a certain multivariate complex Gaussian distribution (an introduction). *The Annals of Mathematical Statistics*, 34(1):152–177, 1963. DOI: 10.1214/aoms/1177704250 27

[93] Todd K. Moon. The expectation-maximization algorithm. *IEEE Signal Processing Magazine*, 13(6):47–60, 1996. DOI: 10.1109/79.543975 27

[94] Jongmin Lee, Michael Stanley, Andreas Spanias, and Cihan Tepedelenlioglu. Integrating machine learning in embedded sensor systems for internet-of-things applications. *IEEE International Symposium on Signal Processing and Information Technology (ISSPIT)*, pages 290–294, 2016. DOI: 10.1109/isspit.2016.7886051 27, 60

[95] Jie Cai, Jiawei Luo, Shulin Wang, and Sheng Yang. Feature selection in machine learning: A new perspective. *Neurocomputing*, 300:70–79, 2018. DOI: 10.1016/j.neucom.2017.11.077 28

[96] Isabelle Guyon, Steve Gunn, Masoud Nikravesh, and Lofti A. Zadeh. *Feature Extraction: Foundations and Applications*, vol. 207. Springer, 2008. DOI: 10.1007/978-3-540-35488-8 28

[97] Yoshua Bengio, Aaron Courville, and Pascal Vincent. Representation learning: A review and new perspectives. *IEEE Transactions on Pattern Analysis and Machine Intelligence*, 35(8):1798–1828, 2013. DOI: 10.1109/tpami.2013.50 28, 54

[98] Karl Pearson. LIII. On lines and planes of closest fit to systems of points in space. *The London, Edinburgh, and Dublin Philosophical Magazine and Journal of Science*, 2(11):559–572, 1901. DOI: 10.1080/14786440109462720 28

[99] Aapo Hyvärinen and Erkki Oja. Independent component analysis: Algorithms and applications. *Neural Networks*, 13(4-5):411–430, 2000. DOI: 10.1016/s0893-6080(00)00026-5 29

[100] James V. Stone. *Independent Component Analysis: A Tutorial Introduction*. The MIT Press, 2004. DOI: 10.7551/mitpress/3717.001.0001 29

[101] Adelbert W. Bronkhorst. The cocktail party phenomenon: A review of research on speech intelligibility in multiple-talker conditions. *Acta Acustica United with Acustica*, 86(1):117–128, 2000. 30

[102] Laurens Van der Maaten and Geoffrey Hinton. Visualizing data using t-SNE. *Journal of Machine Learning Research*, 9(11):2008. 30

[103] Laurens Van Der Maaten. Learning a parametric embedding by preserving local structure. *Artificial Intelligence and Statistics*, pages 384–391, PMLR, 2009. 30

[104] Solomon Kullback. *Information Theory and Statistics*. Courier Corporation, 1997. 30

[105] Alan Wisler, Visar Berisha, Andreas Spanias, and Alfred O. Hero. Direct estimation of density functionals using a polynomial basis. *IEEE Transactions on Signal Processing*, 66(3):558–572, 2017. DOI: 10.1109/tsp.2017.2775587 30

[106] Visar Berisha, Alan Wisler, Alfred O. Hero, and Andreas Spanias. Empirically estimable classification bounds based on a nonparametric divergence measure. *IEEE Transactions on Signal Processing*, 64(3):580–591, 2015. DOI: 10.1109/tsp.2015.2477805 30

[107] Nizar Grira, Michel Crucianu, and Nozha Boujemaa. Unsupervised and semi-supervised clustering: A brief survey. *A Review of Machine Learning Techniques for Processing Multimedia Content*, 1:9–16, 2004. 33

[108] Xiaojin Zhu and Andrew B. Goldberg. Introduction to semi-supervised learning. *Synthesis Lectures on Artificial Intelligence and Machine Learning*, 3(1):1–130, 2009. DOI: 10.2200/s00196ed1v01y200906aim006 33

[109] Olivier Chapelle, Bernhard Scholkopf, and Alexander Zien. Semi-supervised learning (Chapelle, O. et al., Eds., 2006) [book reviews]. *IEEE Transactions on Neural Networks*, 20(3):542–542, 2009. DOI: 10.1109/TNN.2009.2015974 33

[110] Simen Skaret Karlsen. Automated front detection-using computer vision and machine learning to explore a new direction in automated weather forecasting. Master's thesis, The University of Bergen, 2017. 33, 61

[111] Alex Graves, Navdeep Jaitly, and Abdel-Rahman Mohamed. Hybrid speech recognition with deep bidirectional LSTM. *IEEE Workshop on Automatic Speech Recognition and Understanding*, pages 273–278, 2013. DOI: 10.1109/asru.2013.6707742 33, 54

[112] Amarnag Subramanya and Partha Pratim Talukdar. Graph-based semi-supervised learning. *Synthesis Lectures on Artificial Intelligence and Machine Learning*, 8(4):1–125, 2014. DOI: 10.2200/s00590ed1v01y201408aim029 34

[113] David I. Shuman, Sunil K. Narang, Pascal Frossard, Antonio Ortega, and Pierre Vandergheynst. The emerging field of signal processing on graphs: Extending high-dimensional data analysis to networks and other irregular domains. *IEEE Signal Processing Magazine*, 30(3):83–98, 2013. DOI: 10.1109/msp.2012.2235192 35

[114] Antonio Ortega, Pascal Frossard, Jelena Kovačević, José M. F. Moura, and Pierre Vandergheynst. Graph signal processing: Overview, challenges, and applications. *Proc. of the IEEE*, 106(5):808–828, 2018. DOI: 10.1109/jproc.2018.2820126 35

[115] Zheng-Jun Zha, Tao Mei, Jingdong Wang, Zengfu Wang, and Xian-Sheng Hua. Graph-based semi-supervised learning with multiple labels. *Journal of Visual Communication and Image Representation*, 20(2):97–103, 2009. DOI: 10.1016/j.jvcir.2008.11.009 35

[116] Hisashi Kashima, Tsuyoshi Kato, Yoshihiro Yamanishi, Masashi Sugiyama, and Koji Tsuda. Link propagation: A fast semi-supervised learning algorithm for link prediction. *Proc. of the SIAM International Conference on Data Mining*, pages 1100–1111, 2009. DOI: 10.1137/1.9781611972795.94 35

[117] Muhan Zhang and Yixin Chen. Link prediction based on graph neural networks. *Advances in Neural Information Processing Systems*, 31:5165–5175, 2018. 35

[118] Xiaoke Ma, Lin Gao, Xuerong Yong, and Lidong Fu. Semi-supervised clustering algorithm for community structure detection in complex networks. *Physica A: Statistical Mechanics and its Applications*, 389(1):187–197, 2010. DOI: 10.1016/j.physa.2009.09.018 35

[119] Leman Akoglu, Hanghang Tong, and Danai Koutra. Graph based anomaly detection and description: A survey. *Data Mining and Knowledge Discovery*, 29(3):626–688, 2015. DOI: 10.1007/s10618-014-0365-y 35

[120] Rushil Anirudh, Jayaraman J. Thiagarajan, Rahul Sridhar, and Peer-Timo Bremer. Margin: Uncovering deep neural networks using graph signal analysis. *ArXiv Preprint ArXiv:1711.05407*, 2017. DOI: 10.3389/fdata.2021.589417 36

[121] Sandro Cavallari, Vincent W. Zheng, Hongyun Cai, Kevin Chen-Chuan Chang, and Erik Cambria. Learning community embedding with community detection and node embedding on graphs. *Proc. of the ACM on Conference on Information and Knowledge Management*, pages 377–386, 2017. DOI: 10.1145/3132847.3132925 36

[122] Mason A. Porter, Jukka-Pekka Onnela, and Peter J. Mucha. Communities in networks. *Notices of the AMS*, 56(9):1082–1097, 2009. 36

[123] Bryan Perozzi, Rami Al-Rfou, and Steven Skiena. DeepWalk: Online learning of social representations. *Proc. of the 20th ACM SIGKDD International Conference on Knowledge Discovery and Data Mining—KDD'14*, pages 701–710, ACM Press, New York, 2014. DOI: 10.1145/2623330.2623732 38

[124] Masayuki Karasuyama and Hiroshi Mamitsuka. Manifold-based similarity adaptation for label propagation. *Advances in Neural Information Processing Systems*, 26:1547–1555, 2013. 37

[125] Rie Kubota Ando and Tong Zhang. Learning on graph with Laplacian regularization. *Advances in Neural Information Processing Systems*, 19:25, 2007. DOI: 10.7551/mitpress/7503.003.0009 37

[126] Amr Ahmed, Nino Shervashidze, Shravan Narayanamurthy, Vanja Josifovski, and Alexander J. Smola. Distributed large-scale natural graph factorization. *Proc. of the 22nd International Conference on World Wide Web*, pages 37–48, ACM, 2013. DOI: 10.1145/2488388.2488393 37

[127] Mark E. J. Newman. Finding community structure in networks using the eigenvectors of matrices. *Physical Review E*, 74(3):036104, 2006. DOI: 10.1103/physreve.74.036104 37

[128] Mingming Chen, Konstantin Kuzmin, and Boleslaw K. Szymanski. Community detection via maximization of modularity and its variants. *IEEE Transactions on Computational Social Systems*, 1(1):46–65, 2014. DOI: 10.1109/tcss.2014.2307458 37

[129] Zellig Harris. Distributional structure. *Word*, 10(2–3):146–162, 1954. DOI: 10.1080/00437956.1954.11659520 38

[130] Aditya Grover and Jure Leskovec. node2vec. *Proc. of the 22nd ACM SIGKDD International Conference on Knowledge Discovery and Data Mining*, ACM, August 2016. DOI: 10.1145/2939672.2939754 38

[131] Thomas N. Kipf and Max Welling. Semi-supervised classification with graph convolutional networks. *ArXiv Preprint ArXiv:1609.02907*, 2016. 38

[132] Michaël Defferrard, Xavier Bresson, and Pierre Vandergheynst. Convolutional neural networks on graphs with fast localized spectral filtering. *Advances in Neural Information Processing Systems*, pages 3844–3852, 2016. 38

[133] Charles Elkan and Keith Noto. Learning classifiers from only positive and unlabeled data. *Proc. of the 14th ACM SIGKDD International Conference on Knowledge Discovery and Data Mining*, pages 213–220, 2008. DOI: 10.1145/1401890.1401920 39

[134] Azam Kaboutari, J. Bagherzadeh, and F. Kheradmand. An evaluation of two-step techniques for positive-unlabeled learning in text classification. *International Journal of Computer Applications in Technology Research*, 3:592–594, 2014. DOI: 10.7753/ijcatr0309.1012 40

[135] Francois Denis, Remi Gilleron, and Marc Tommasi. Text classification from positive and unlabeled examples. *Proc. of the 9th International Conference on Information Processing and Management of Uncertainty in Knowledge-Based Systems, IPMU'02*, pages 1927–1934, 2002. 40

[136] Ryuichi Kiryo, Gang Niu, Marthinus C. du Plessis, and Masashi Sugiyama. Positive-unlabeled learning with non-negative risk estimator. *ArXiv Preprint ArXiv:1703.00593*, 2017. 40

[137] Kristen Jaskie and Andreas Spanias. Positive and unlabeled learning algorithms and applications: A survey. *10th International Conference on Information, Intelligence, Systems and Applications (IEEE IISA)*, pages 1–8, July, 2019. DOI: 10.1109/iisa.2019.8900698 40

[138] Jessa Bekker and Jesse Davis. Learning from positive and unlabeled data: A survey. *Machine Learning*, 109(4):719–760, April 2020. DOI: 10.1007/s10994-020-05877-5 40

[139] Gang Niu, Marthinus Christoffel du Plessis, Tomoya Sakai, Yao Ma, and Masashi Sugiyama. Theoretical comparisons of positive-unlabeled learning against positive-negative learning. *ArXiv Preprint ArXiv:1603.03130*, 2016. 40

[140] Peng Yang, Xiao-Li Li, Jian-Ping Mei, Chee-Keong Kwoh, and See-Kiong Ng. Positive-unlabeled learning for disease gene identification. *Bioinformatics*, 28(20):2640–2647, 2012. DOI: 10.1093/bioinformatics/bts504 40

[141] Xiao-Li Li, Philip S. Yu, Bing Liu, and See-Kiong Ng. Positive unlabeled learning for data stream classification. *Proc. of the SIAM International Conference on Data Mining*, pages 259–270, 2009. DOI: 10.1137/1.9781611972795.23 40

[142] Kristen Jaskie, Dominique Smith, and Andreas Spanias. Deep learning networks for vectorized energy load forecasting. *11th International Conference on Information, Intelligence, Systems and Applications (IISA)*, pages 1–7, 2020. DOI: 10.1109/iisa50023.2020.9284364 40, 68

[143] Li Deng and Dong Yu. Deep learning: Methods and applications. *Foundations and Trends in Signal Processing*, 7(3–4):197–387, 2014. DOI: 10.1561/2000000039 43

[144] Dong Yu and Li Deng. Deep learning and its applications to signal and information processing [exploratory DSP]. *IEEE Signal Processing Magazine*, 28(1):145–154, 2010. DOI: 10.1109/msp.2010.939038 43

[145] Athanasios Voulodimos, Nikolaos Doulamis, Anastasios Doulamis, and Eftychios Protopapadakis. Deep learning for computer vision: A brief review. *Computational Intelligence and Neuroscience*, 2018. DOI: 10.1155/2018/7068349 43

[146] Hendrik Purwins, Bo Li, Tuomas Virtanen, Jan Schlüter, Shuo-Yiin Chang, and Tara Sainath. Deep learning for audio signal processing. *IEEE Journal of Selected Topics in Signal Processing*, 13(2):206–219, 2019. DOI: 10.1109/jstsp.2019.2908700 43

[147] Li Deng and Yang Liu. *Deep Learning in Natural Language Processing*. Springer, 2018. DOI: 10.1007/978-981-10-5209-5_10 43

[148] Tom Young, Devamanyu Hazarika, Soujanya Poria, and Erik Cambria. Recent trends in deep learning based natural language processing. *IEEE Computational Intelligence Magazine*, 13(3):55–75, 2018. DOI: 10.1109/mci.2018.2840738 43

[149] Matt W. Gardner and S. R. Dorling. Artificial neural networks (the multilayer perceptron)—a review of applications in the atmospheric sciences. *Atmospheric Environment*, 32(14–15):2627–2636, 1998. DOI: 10.1016/s1352-2310(97)00447-0 44

[150] Sebastian Ruder. An overview of gradient descent optimization algorithms. *ArXiv Preprint ArXiv:1609.04747*, 2016. 45

[151] Henry Leung and Simon Haykin. The complex backpropagation algorithm. *IEEE Transactions on Signal Processing*, 39(9):2101–2104, 1991. DOI: 10.1109/78.134446 46

[152] Raul Rojas. The backpropagation algorithm. *Neural Networks*, pages 149–182, Springer, 1996. DOI: 10.1007/978-3-642-61068-4_7 46

[153] Sagar Sharma. Activation functions in neural networks. *Towards Data Science*, 6, 2017. DOI: 10.33564/ijeast.2020.v04i12.054 47

[154] Sergey Ioffe and Christian Szegedy. Batch normalization: Accelerating deep network training by reducing internal covariate shift. *International Conference on Machine Learning*, pages 448–456, PMLR, 2015. 48

[155] Tim Salimans and Diederik P. Kingma. Weight normalization: A simple reparameterization to accelerate training of deep neural networks. *ArXiv Preprint ArXiv:1602.07868*, 2016. 49

[156] Jimmy Lei Ba, Jamie Ryan Kiros, and Geoffrey E. Hinton. Layer normalization. *ArXiv Preprint ArXiv:1607.06450*, 2016. 49

[157] Nitish Srivastava, Geoffrey Hinton, Alex Krizhevsky, Ilya Sutskever, and Ruslan Salakhutdinov. Dropout: A simple way to prevent neural networks from overfitting. *The Journal of Machine Learning Research*, 15(1):1929–1958, 2014. 49

[158] Connor Shorten and Taghi M. Khoshgoftaar. A survey on image data augmentation for deep learning. *Journal of Big Data*, 6(1):1–48, 2019. DOI: 10.1186/s40537-019-0197-0 49

[159] Robert Moore and John DeNero. L1 and l2 regularization for multiclass hinge loss models. *Symposium on Machine Learning in Speech and Language Processing*, 2011. 49

[160] Bernhard Schölkopf, Alexander J. Smola, Francis Bach, et al. *Learning with Kernels: Support Vector Machines, Regularization, Optimization, and Beyond*. MIT Press, 2002. DOI: 10.7551/mitpress/4175.001.0001 49

[161] Thomas G. Dietterich et al. Ensemble learning. *The Handbook of Brain Theory and Neural Networks*, 2:110–125, 2002. 50

[162] X. Glorot and Y. Bengio. Very deep convolutional networks for large-scale image recognition. *Proc. of the 13th International Conference on Artificial Intelligence and Statistics*, 2010. 50

[163] Alex Krizhevsky, Ilya Sutskever, and Geoffrey E. Hinton. ImageNet classification with deep convolutional neural networks. *Advances in Neural Information Processing Systems*, 25:1097–1105, 2012. DOI: 10.1145/3065386 52

[164] Karen Simonyan and Andrew Zisserman. Very deep convolutional networks for large-scale image recognition. *ArXiv Preprint ArXiv:1409.1556*, 2014. 52

[165] Christian Szegedy, Sergey Ioffe, Vincent Vanhoucke, and Alexander A. Alemi. Inception-v4, inception-resnet and the impact of residual connections on learning. *31st AAAI Conference on Artificial Intelligence*, 2017. 52

[166] Kaiming He, Xiangyu Zhang, Shaoqing Ren, and Jian Sun. Deep residual learning for image recognition. *Proc. of the IEEE Conference on Computer Vision and Pattern Recognition*, pages 770–778, 2016. DOI: 10.1109/cvpr.2016.90 52

[167] David E. Rumelhart, Geoffrey E. Hinton, and Ronald J. Williams. Learning internal representations by error propagation. *Technical Report*, California University San Diego La Jolla Institute for Cognitive Science, 1985. DOI: 10.21236/ada164453 53

[168] Alex Sherstinsky. Fundamentals of recurrent neural network (RNN) and long short-term memory (LSTM) network. *Physica D: Nonlinear Phenomena*, 404:132306, March 2020. DOI: 10.1016/j.physd.2019.132306 53

[169] Paul J. Werbos. Backpropagation through time: What it does and how to do it. *Proc. of the IEEE*, 78(10):1550–1560, 1990. DOI: 10.1109/5.58337 54

[170] Klaus Greff, Rupesh K. Srivastava, Jan Koutník, Bas R. Steunebrink, and Jürgen Schmidhuber. LSTM: A search space odyssey. *IEEE Transactions on Neural Networks and Learning Systems*, 28(10):2222–2232, 2016. DOI: 10.1109/tnnls.2016.2582924 54

[171] Mark A. Kramer. Nonlinear principal component analysis using autoassociative neural networks. *AICHE Journal*, 37(2):233–243, 1991. DOI: 10.1002/aic.690370209 55

[172] Alireza Makhzani and Brendan Frey. K-sparse autoencoders. *ArXiv Preprint ArXiv:1312.5663*, 2013. 55

[173] Pascal Vincent, Hugo Larochelle, Isabelle Lajoie, Yoshua Bengio, Pierre-Antoine Manzagol, and Léon Bottou. Stacked denoising autoencoders: Learning useful representations in a deep network with a local denoising criterion. *Journal of Machine Learning Research*, 11(12):2010. 55

[174] Alec Radford, Luke Metz, and Soumith Chintala. Unsupervised representation learning with deep convolutional generative adversarial networks, 2016. 55

[175] Martin Arjovsky, Soumith Chintala, and Léon Bottou. Wasserstein gan, 2017. 56

[176] Ishaan Gulrajani, Faruk Ahmed, Martin Arjovsky, Vincent Dumoulin, and Aaron Courville. Improved training of Wasserstein gans, 2017. 56

[177] Mehdi Mirza and Simon Osindero. Conditional generative adversarial nets, 2014. 56

[178] Yujun Shen, Jinjin Gu, Xiaoou Tang, and Bolei Zhou. Interpreting the latent space of gans for semantic face editing, 2020. DOI: 10.1109/cvpr42600.2020.00926 56

[179] Xi Chen, Yan Duan, Rein Houthooft, John Schulman, Ilya Sutskever, and Pieter Abbeel. Infogan: Interpretable representation learning by information maximizing generative adversarial nets, 2016. 56

[180] Tero Karras, Samuli Laine, and Timo Aila. A style-based generator architecture for generative adversarial networks. *Proc. of the IEEE/CVF Conference on Computer Vision and Pattern Recognition*, pages 4401–4410, 2019. DOI: 10.1109/cvpr.2019.00453 56

[181] Z. Yu and J. J. P. Tsai. A framework of machine learning based intrusion detection for wireless sensor networks. *IEEE International Conference on Sensor Networks, Ubiquitous, and Trustworthy Computing (SUTC)*, pages 272–279, 2008. DOI: 10.1109/sutc.2008.39 59

[182] Shui-Bo Zheng, Zheng-Zhi Han, Hou-Jun Tang, and Yong Zhang. Application of support vector machines to sensor fault diagnosis in ESP system. *Proc. of International Conference on Machine Learning and Cybernetics (IEEE Cat. no. 04EX826)*, 6:3334–3338, 2004. DOI: 10.1109/icmlc.2004.1380354 60

[183] R. Huan, Q. Chen, K. Mao, and Y. Pan. A three-dimension localization algorithm for wireless sensor network nodes based on SVM. *The International Conference on Green Circuits and Systems*, pages 651–654, 2010. DOI: 10.1109/icgcs.2010.5542981 60

[184] Michael Stanley and Jongmin Lee. Sensor analysis for the internet of things. *Synthesis Lectures on Algorithms and Software in Engineering*, 9(1):1–137, 2018. DOI: 10.2200/s00827ed1v01201802ase017 60

[185] Y. Wang, A. Yang, X. Chen, P. Wang, Y. Wang, and H. Yang. A deep learning approach for blind drift calibration of sensor networks. *IEEE Sensors Journal*, 17(13):4158–4171, 2017. DOI: 10.1109/jsen.2017.2703885 60

[186] T. Zebin, P. J. Scully, and K. B. Ozanyan. Human activity recognition with inertial sensors using a deep learning approach. *IEEE Sensors*, pages 1–3, 2016. DOI: 10.1109/icsens.2016.7808590 60

[187] Honggui Li and Maria Trocan. Deep learning of smartphone sensor data for personal health assistance. *Microelectronics Journal*, 88:164–172, 2019. DOI: 10.1016/j.mejo.2018.01.015 60

[188] Yu Han and Y. H. Song. Condition monitoring techniques for electrical equipment-a literature survey. *IEEE Transactions on Power Delivery*, 18(1):4–13, 2003. DOI: 10.1109/tpwrd.2002.801425 60

[189] Philip A. Higgs, Rob Parkin, Mike Jackson, Amin Al-Habaibeh, Farbod Zorriassatine, and Jo Coy. A survey on condition monitoring systems in industry. *Engineering Systems Design and Analysis*, 41758:163–178, 2004. DOI: 10.1115/esda2004-58216 60

[190] G. K. Singh et al. Induction machine drive condition monitoring and diagnostic research—a survey. *Electric Power Systems Research*, 64(2):145–158, 2003. DOI: 10.1016/s0378-7796(02)00172-4 60

[191] M. Craig, T. J. Harvey, R. J. K. Wood, K. Masuda, M. Kawabata, and H. E. G. Powrie. Advanced condition monitoring of tapered roller bearings, part 1. *Tribology International*, 42(11–12):1846–1856, 2009. DOI: 10.1016/j.triboint.2009.04.033 60

[192] Wei Qiao and Dingguo Lu. A survey on wind turbine condition monitoring and fault diagnosis—part I: Components and subsystems. *IEEE Transactions on Industrial Electronics*, 62(10):6536–6545, 2015. DOI: 10.1109/tie.2015.2422112 60

[193] Christopher J. Crabtree, Donatella Zappalá, and Peter J. Tavner. Survey of commercially available condition monitoring systems for wind turbines, 2014. 60

[194] Bin Lu, Yaoyu Li, Xin Wu, and Zhongzhou Yang. A review of recent advances in wind turbine condition monitoring and fault diagnosis. *IEEE Power Electronics and Machines in Wind Applications*, pages 1–7, 2009. DOI: 10.1109/pemwa.2009.5208325 60

[195] Sorin Berbente, Irina-Carmen Andrei, Gabriela Stroe, and Mihaela-Luminita Costea. Topical issues in aircraft health management with applications to jet engines. *INCAS Bulletin*, 12(1):13–26, 2020. DOI: 10.13111/2066-8201.2020.12.1.2 60

[196] Ahmed EB Abu-Elanien and M. M. A. Salama. Survey on the transformer condition monitoring. *Large Engineering Systems Conference on Power Engineering*, pages 187–191, IEEE, 2007. DOI: 10.1109/lescpe.2007.4437376 60

[197] M Zekveld and Gerhard P. Hancke. Vibration condition monitoring using machine learning. *IECON 44th Annual Conference of the IEEE Industrial Electronics Society*, pages 4742–4747, 2018. DOI: 10.1109/iecon.2018.8591167 60

[198] Xiaolin Meng. Real-time deformation monitoring of bridges using GPS/accelerometers. Ph.D. thesis, University of Nottingham Nottingham, UK, 2002. 60

[199] W. S. Chan, You Lin Xu, X. L. Ding, and W. J. Dai. An integrated GPS—accelerometer data processing technique for structural deformation monitoring. *Journal of Geodesy*, 80(12):705–719, 2006. DOI: 10.1007/s00190-006-0092-2 60

[200] Parijata Majumdar and Jhunu Debbarma. Detection and removal of infrared (IR) image noise patterns: An experimental case study of knee thermograms. *International Journal of Computational Intelligence and IoT*, 1(2):2018. 60

[201] R. Collacott. *Mechanical Fault Diagnosis and Condition Monitoring*. Springer Science & Business Media, 2012. DOI: 10.1007/978-94-009-5723-7 60

[202] Stephanie Renee Debats. Mapping Sub-Saharan African agriculture in high-resolution satellite imagery with computer vision and machine learning. Ph.D. thesis, Princeton University, 2017. 61

[203] H. Durmuş, E. O. Güneş, and M. Kırcı. Disease detection on the leaves of the tomato plants by using deep learning. *6th International Conference on Agro-Geoinformatics*, pages 1–5, 2017. DOI: 10.1109/agro-geoinformatics.2017.8047016 61

[204] Gu Yunchao and Yang Jiayao. Application of computer vision and deep learning in breast cancer assisted diagnosis. *Proc. of the 3rd International Conference on Machine Learning and Soft Computing*, pages 186–191, 2019. DOI: 10.1145/3310986.3311010 61

[205] Shervin Minaee, Rahele Kafieh, Milan Sonka, Shakib Yazdani, and Ghazaleh Jamalipour Soufi. Deep-covid: Predicting covid-19 from chest x-ray images using deep transfer learning. *ArXiv Preprint ArXiv:2004.09363*, 2020. DOI: 10.1016/j.media.2020.101794 61

[206] B. T. Nugraha, S. Su, and Fahmizal. Towards self-driving car using convolutional neural network and road lane detector. *2nd International Conference on Automation, Cognitive Science, Optics, Micro Electro–Mechanical System, and Information Technology (ICACOMIT)*, pages 65–69, 2017. DOI: 10.1109/icacomit.2017.8253388 61

[207] Satoshi Iizuka, Edgar Simo-Serra, and Hiroshi Ishikawa. Let there be color! joint end-to-end learning of global and local image priors for automatic image colorization with simultaneous classification. *ACM Transactions on Graphics (ToG)*, 35(4):1–11, 2016. DOI: 10.1145/2897824.2925974 61

[208] L. A. Gatys, A. S. Ecker, and M. Bethge. Image style transfer using convolutional neural networks. *IEEE Conference on Computer Vision and Pattern Recognition (CVPR)*, pages 2414–2423, 2016. DOI: 10.1109/cvpr.2016.265 61

[209] Zezhou Cheng, Qingxiong Yang, and Bin Sheng. Deep colorization. *CoRR*, 2016. DOI: 10.1109/iccv.2015.55 61

[210] C. Amritkar and V. Jabade. Image caption generation using deep learning technique. *4th International Conference on Computing Communication Control and Automation (ICCUBEA)*, pages 1–4, 2018. DOI: 10.1109/iccubea.2018.8697360 61

[211] M. D. Zakir Hossain, Ferdous Sohel, Mohd Fairuz Shiratuddin, and Hamid Laga. A comprehensive survey of deep learning for image captioning. *ACM Computing Surveys (CSUR)*, 51(6):1–36, 2019. DOI: 10.1145/3295748 61

[212] Szilárd Aradi. Survey of deep reinforcement learning for motion planning of autonomous vehicles. *IEEE Transactions on Intelligent Transportation Systems*, 2020. DOI: 10.1109/tits.2020.3024655 62

[213] Moritz Menze and Andreas Geiger. Object scene flow for autonomous vehicles. *Proc. of the IEEE Conference on Computer Vision and Pattern Recognition*, pages 3061–3070, 2015. DOI: 10.1109/cvpr.2015.7298925 62

[214] Shashi D. Buluswar and Bruce A. Draper. Color machine vision for autonomous vehicles. *Engineering Applications of Artificial Intelligence*, 11(2):245–256, 1998. DOI: 10.1016/s0952-1976(97)00079-1 62

[215] Joel Janai, Fatma Güney, Aseem Behl, Andreas Geiger, et al. Computer vision for autonomous vehicles: Problems, datasets and state of the art. *Foundations and Trends® in Computer Graphics and Vision*, 12(1–3):1–308, 2020. DOI: 10.1561/0600000079 62

[216] Vipin Kumar Kukkala, Jordan Tunnell, Sudeep Pasricha, and Thomas Bradley. Advanced driver-assistance systems: A path toward autonomous vehicles. *IEEE Consumer Electronics Magazine*, 7(5):18–25, 2018. DOI: 10.1109/mce.2018.2828440 62

[217] Akshay Rangesh and Mohan Manubhai Trivedi. No blind spots: Full-surround multi-object tracking for autonomous vehicles using cameras and lidars. *IEEE Transactions on Intelligent Vehicles*, 4(4):588–599, 2019. DOI: 10.1109/tiv.2019.2938110 62

[218] Jiadai Wang, Jiajia Liu, and Nei Kato. Networking and communications in autonomous driving: A survey. *IEEE Communications Surveys and Tutorials*, 21(2):1243–1274, 2018. DOI: 10.1109/comst.2018.2888904 62

[219] Haixia Peng, Dazhou Li, Khadige Abboud, Haibo Zhou, Hai Zhao, Weihua Zhuang, and Xuemin Shen. Performance analysis of IEEE 802.11p DCF for multiplatooning communications with autonomous vehicles. *IEEE Transactions on Vehicular Technology*, 66(3):2485–2498, 2016. DOI: 10.1109/tvt.2016.2571696 62

[220] Rafael Molina-Masegosa and Javier Gozalvez. LTE-V for sidelink 5G V2X vehicular communications: A new 5G technology for short-range vehicle-to-everything communications. *IEEE Vehicular Technology Magazine*, 12(4):30–39, 2017. DOI: 10.1109/mvt.2017.2752798 62

[221] Sampo Kuutti, Saber Fallah, Konstantinos Katsaros, Mehrdad Dianati, Francis Mccullough, and Alexandros Mouzakitis. A survey of the state-of-the-art localization techniques and their potentials for autonomous vehicle applications. *IEEE Internet of Things Journal*, 5(2):829–846, 2018. DOI: 10.1109/jiot.2018.2812300 62

[222] Parham M. Kebria, Abbas Khosravi, Syed Moshfeq Salaken, and Saeid Nahavandi. Deep imitation learning for autonomous vehicles based on convolutional neural networks. *IEEE/CAA Journal of Automatica Sinica*, 7(1):82–95, 2020. DOI: 10.1109/jas.2019.1911825 62

[223] Erik G. Larsson, Ove Edfors, Fredrik Tufvesson, and Thomas L. Marzetta. Massive MIMO for next generation wireless systems. *IEEE Communications Magazine*, 52(2):186–195, 2014. DOI: 10.1109/mcom.2014.6736761 63

[224] R. C. D. Lamare. MIMO systems: Signal processing challenges and future trends. *URSI Radio Science Bulleting*, 347:8–20. 63

[225] Lu Lu, Geoffrey Ye Li, A. Lee Swindlehurst, Alexei Ashikhmin, and Rui Zhang. An overview of massive MIMO: Benefits and challenges. *IEEE Journal of Selected Topics in Signal Processing*, 8(5):742–758, 2014. DOI: 10.1109/jstsp.2014.2317671 63

[226] Federico Boccardi, Robert W. Heath, Angel Lozano, Thomas L. Marzetta, and Petar Popovski. Five disruptive technology directions for 5G. *IEEE Communications Magazine*, 52(2):74–80, 2014. DOI: 10.1109/mcom.2014.6736746 63

[227] Jayden Booth, Ahmed Ewaisha, Andreas Spanias, and Ahmed Alkhateeb. Deep learning based MIMO channel prediction: An initial proof of concept prototype. *54th Asilomar Conference on Signals, Systems, and Computers*, pages 267–271, IEEE, 2020. DOI: 10.1109/ieeeconf51394.2020.9443515 63

[228] Theodore S. Rappaport, Yunchou Xing, Ojas Kanhere, Shihao Ju, Arjuna Madanayake, Soumyajit Mandal, Ahmed Alkhateeb, and Georgios C. Trichopoulos. Wireless communications and applications above 100 GHZ: Opportunities and challenges for 6G and beyond. *IEEE Access*, 7:78729–78757, 2019. DOI: 10.1109/access.2019.2921522 63

[229] Ahmed Alkhateeb, Geert Leus, and Robert W. Heath. Limited feedback hybrid precoding for multi-user millimeter wave systems. *IEEE Transactions on Wireless Communications*, 14(11):6481–6494, 2015. DOI: 10.1109/twc.2015.2455980 63

[230] Jun Liu, Kai Mei, Xiaochen Zhang, Dongtang Ma, and Jibo Wei. Online extreme learning machine-based channel estimation and equalization for OFDM systems. *IEEE Communications Letters*, 23(7):1276–1279, 2019. DOI: 10.1109/lcomm.2019.2916797 63

[231] Osvaldo Simeone. A very brief introduction to machine learning with applications to communication systems. *IEEE Transactions on Cognitive Communications and Networking*, 4(4):648–664, 2018. DOI: 10.1109/tccn.2018.2881442 63

[232] Muhammad Amjad, Fayaz Akhtar, Mubashir Husain Rehmani, Martin Reisslein, and Tariq Umer. Full-duplex communication in cognitive radio networks: A survey. *IEEE Communications Surveys and Tutorials*, 19(4):2158–2191, 2017. DOI: 10.1109/comst.2017.2718618 63

[233] Yong Niu, Yong Li, Depeng Jin, Li Su, and Athanasios V. Vasilakos. A survey of millimeter wave communications (mmWave) for 5G: Opportunities and challenges. *Wireless Networks*, 21(8):2657–2676, 2015. DOI: 10.1007/s11276-015-0942-z 63

[234] Aleksandar Damnjanovic, Juan Montojo, Yongbin Wei, Tingfang Ji, Tao Luo, Madhavan Vajapeyam, Taesang Yoo, Osok Song, and Durga Malladi. A survey on 3GPP heterogeneous networks. *IEEE Wireless Communications*, 18(3):10–21, 2011. DOI: 10.1109/mwc.2011.5876496 63

[235] Akhil Gupta and Rakesh Kumar Jha. A survey of 5G network: Architecture and emerging technologies. *IEEE Access*, 3:1206–1232, 2015. DOI: 10.1109/access.2015.2461602 63

[236] Maximilian Arnold, Sebastian Dörner, Sebastian Cammerer, Sarah Yan, Jakob Hoydis, and Stephan Ten Brink. Enabling FDD massive MIMO through deep learning-based channel prediction. *ArXiv Preprint ArXiv:1901.03664*, 2019. 63

[237] Yuwen Yang, Feifei Gao, Geoffrey Ye Li, and Mengnan Jian. Deep learning-based downlink channel prediction for FDD massive MIMO system. *IEEE Communications Letters*, 23(11):1994–1998, 2019. DOI: 10.1109/lcomm.2019.2934851 63

[238] Nariman Farsad, Milind Rao, and Andrea Goldsmith. Deep learning for joint source-channel coding of text. *IEEE International Conference on Acoustics, Speech and Signal Processing (ICASSP)*, pages 2326–2330, 2018. DOI: 10.1109/icassp.2018.8461983 63

[239] Anestis Tsakmalis, Symeon Chatzinotas, and Björn Ottersten. Automatic modulation classification for adaptive power control in cognitive satellite communications. *7th Advanced Satellite Multimedia Systems Conference and the 13th Signal Processing for Space Communications Workshop (ASMS/SPSC)*, pages 234–240, IEEE, 2014. DOI: 10.1109/asms-spsc.2014.6934549 63

[240] Q. V. Hu, X. J. Huang, and J. Miao. Exploring a multi-source fusion approach for genomics information retrieval. *IEEE International Conference on Bioinformatics and Biomedicine (BIBM)*, pages 669–672, 2010. DOI: 10.1109/bibm.2010.5706649 64

[241] K. Chen, R. Wang, M. Utiyama, E. Sumita, T. Zhao, M. Yang, and H. Zhao. Towards more diverse input representation for neural machine translation. *IEEE/ACM Transactions on Audio, Speech, and Language Processing*, 28:1586–1597, 2020. DOI: 10.1109/taslp.2020.2996077 64

[242] Y. Woldemariam. Sentiment analysis in a cross-media analysis framework. *IEEE International Conference on Big Data Analysis (ICBDA)*, pages 1–5, 2016. DOI: 10.1109/icbda.2016.7509790 64

[243] Ping-I Chen and Shi-Jen Lin. Automatic keyword prediction using Google similarity distance. *Expert Systems with Applications*, 37(3):1928–1938, 2010. DOI: 10.1016/j.eswa.2009.07.016 64

[244] Tomas Mikolov, Kai Chen, Greg Corrado, and Jeffrey Dean. Efficient estimation of word representations in vector space. *ArXiv Preprint ArXiv:1301.3781*, 2013. 64

[245] Tomas Mikolov, Ilya Sutskever, Kai Chen, Greg S. Corrado, and Jeff Dean. Distributed representations of words and phrases and their compositionality. *Advances in Neural Information Processing Systems*, 26:3111–3119, 2013. 64

[246] Ashish Vaswani, Noam Shazeer, Niki Parmar, Jakob Uszkoreit, Llion Jones, Aidan N. Gomez, Łukasz Kaiser, and Illia Polosukhin. Attention is all you need. *Advances in Neural Information Processing Systems*, pages 5998–6008, 2017. 65

[247] A. S. Spanias. Speech coding: A tutorial review. *Proc. of the IEEE*, 82(10):1541–1582, 1994. DOI: 10.1109/5.326413 65

[248] Andreas Spanias, Ted Painter, and Venkatraman Atti. *Audio Signal Processing and Coding*. John Wiley & Sons, 2006. DOI: 10.1002/0470041978 65

[249] M. Schroeder and B. Atal. Code-excited linear prediction (CELP): High-quality speech at very low bit rates. *ICASSP85. IEEE International Conference on Acoustics, Speech, and Signal Processing*, 10:937–940, 1985. DOI: 10.1109/icassp.1985.1168147 65

[250] Karthikeyan N. Ramamurthy and Andreas S. Spanias. Matlab® software for the code excited linear prediction algorithm: The federal standard-1016. *Synthesis Lectures on Algorithms and Software in Engineering*, 2(1):1–109, 2010. DOI: 10.2200/S00252ED1V01Y201001ASE003 65

[251] P. C. Loizou and A. S. Spanias. Vector quantization of transform components for speech coding at 1200 BPS. *Proc. ICASSP: International Conference on Acoustics, Speech, and Signal Processing*, 1:245–248, 1991. DOI: 10.1109/icassp.1991.150323 65

[252] Visar Berisha, Steven Sandoval, and Julie Liss. Bandwidth extension of speech using perceptual criteria. *Synthesis Lectures on Algorithms and Software in Engineering*, Ed. Andreas Spanias, Morgan & Claypool Publishers, 5(2):1–83, 2013. DOI: 10.2200/s00535ed1v01y201309ase013 65

[253] Visar Berisha and Andreas Spanias. Split-band speech compression based on loudness estimation, March 5 2013. 65

[254] Visar Berisha and Andreas Spanias. Wideband speech recovery using psychoacoustic criteria. *EURASIP Journal on Audio, Speech, and Music Processing*, pages 1–18, 2007. DOI: 10.1155/2007/16816 65

[255] L. Rabiner and B. Juang. An introduction to hidden Markov models. *IEEE ASSP Magazine*, 3(1):4–16, 1986. DOI: 10.1109/massp.1986.1165342 65

[256] P. C. Loizou and A. S. Spanias. High-performance alphabet recognition. *IEEE Transactions on Speech and Audio Processing*, 4(6):430–445, 1996. DOI: 10.1109/89.544528 65

[257] Aaron van den Oord, Sander Dieleman, Heiga Zen, Karen Simonyan, Oriol Vinyals, Alex Graves, Nal Kalchbrenner, Andrew Senior, and Koray Kavukcuoglu. WaveNet: A generative model for raw audio. *ArXiv Preprint ArXiv:1609.03499*, 2016. 65

[258] Huan Song, Megan Willi, Jayaraman J. Thiagarajan, Visar Berisha, and Andreas Spanias. Triplet network with attention for speaker diarization. *ArXiv Preprint ArXiv:1808.01535*, 2018. DOI: 10.21437/interspeech.2018-2305 65

[259] Vivek Sivaraman Narayanaswamy, Jayaraman J. Thiagarajan, Huan Song, and Andreas Spanias. Designing an effective metric learning pipeline for speaker diarization. *IEEE International Conference on Acoustics, Speech and Signal Processing (ICASSP)*, pages 5806–5810, 2019. DOI: 10.1109/icassp.2019.8682255 65

[260] H. Hu, M. Xu, and W. Wu. GMM supervector based SVM with spectral features for speech emotion recognition. *IEEE International Conference on Acoustics, Speech and Signal Processing—ICASSP*, 4:IV-413–IV-416, 2007. DOI: 10.1109/icassp.2007.366937 65

[261] Mohit Shah, Ming Tu, Visar Berisha, Chaitali Chakrabarti, and Andreas Spanias. Articulation constrained learning with application to speech emotion recognition. *EURASIP Journal on Audio, Speech, and Music Processing*, (1):1–17, 2019. DOI: 10.1186/s13636-019-0157-9 65

[262] Avery Wang. The shazam music recognition service. *Communications of the ACM*, 49(8):44–48, 2006. DOI: 10.1145/1145287.1145312 66

[263] Roger B. Dannenberg, William P. Birmingham, Bryan Pardo, Ning Hu, Colin Meek, and George Tzanetakis. A comparative evaluation of search techniques for query-by-humming using the musart testbed. *Journal of the American Society for Information Science and Technology*, 58(5):687–701, 2007. DOI: 10.1002/asi.20532 66

[264] Gordon Wichern, Jiachen Xue, Harvey Thornburg, Brandon Mechtley, and Andreas Spanias. Segmentation, indexing, and retrieval for environmental and natural sounds. *IEEE Transactions on Audio, Speech, and Language Processing*, 18(3):688–707, 2010. DOI: 10.1109/tasl.2010.2041384 66

[265] Pablo Gainza, Freyr Sverrisson, Frederico Monti, Emanuele Rodola, D. Boscaini, M. M. Bronstein, and B. E. Correia. Deciphering interaction fingerprints from protein molecular surfaces using geometric deep learning. *Nature Methods*, 17(2):184–192, 2020. DOI: 10.1038/s41592-019-0666-6 66

[266] Andrew W. Senior, Richard Evans, John Jumper, James Kirkpatrick, Laurent Sifre, Tim Green, Chongli Qin, August Žídek, Alexander W. R. Nelson, Alex Bridgland, et al. Improved protein structure prediction using potentials from deep learning. *Nature*, 577(7792):706–710, 2020. DOI: 10.1038/s41586-019-1923-7 66

[267] Jonathan M. Stokes, Kevin Yang, Kyle Swanson, Wengong Jin, Andres Cubillos-Ruiz, Nina M. Donghia, Craig R. MacNair, Shawn French, Lindsey A. Carfrae, Zohar Bloom-Ackerman, et al. A deep learning approach to antibiotic discovery. *Cell*, 180(4):688–702, 2020. DOI: 10.1016/j.cell.2020.01.021 66

[268] Pete Warden and Daniel Situnayake. *TinyML: Machine Learning with TensorFlow Lite on Arduino and Ultra-Low-Power Microcontrollers*. O'Reilly Media, Inc., 2019. 66

[269] M. G. Murshed, Christopher Murphy, Daqing Hou, Nazar Khan, Ganesh Anantha-narayanan, and Faraz Hussain. Machine learning at the network edge: A survey. *ArXiv Preprint ArXiv:1908.00080*, 2019. DOI: 10.1145/3469029 66

[270] Thang Le Duc, Rafael García Leiva, Paolo Casari, and Per-Olov Östberg. Machine learning methods for reliable resource provisioning in edge-cloud computing: A survey. *ACM Computing Surveys (CSUR)*, 52(5):1–39, 2019. DOI: 10.1145/3341145 66

[271] C. Vuppalapati, A. Ilapakurti, S. Kedari, J. Vuppalapati, S. Kedari, and R. Vuppalapati. Democratization of AI, albeit constrained IoT devices tiny ML, for creating a sustainable food future. *3rd International Conference on Information and Computer Technologies (ICICT)*, pages 525–530, 2020. DOI: 10.1109/icict50521.2020.00089 66

[272] TinyML Foundation, https://www.tinyml.org/ 67

[273] TinyML EMEA Forum, https://www.tinyml.org/event/emea-2021/ 67

[274] Varun Gulshan, Lily Peng, Marc Coram, Martin C. Stumpe, Derek Wu, Arunacha-lam Narayanaswamy, Subhashini Venugopalan, Kasumi Widner, Tom Madams, Jorge Cuadros, et al. Development and validation of a deep learning algorithm for detection of diabetic retinopathy in retinal fundus photographs. *Jama*, 316(22):2402–2410, 2016. DOI: 10.1001/jama.2016.17216 67

[275] Taylor Kubota. Deep learning algorithm does as well as dermatologists in identifying skin cancer. Online, January 2017. 67

[276] Mahendra Awale, Finton Sirockin, Nikolaus Stiefl, and Jean-Louis Reymond. Drug analogs from fragment-based long short-term memory generative neural networks. *Journal of Chemical Information and Modeling*, 59(4):1347–1356, 2019. DOI: 10.1021/acs.jcim.8b00902 67

[277] Parminder S. Reel, Smarti Reel, Ewan Pearson, Emanuele Trucco, and Emily Jefferson. Using machine learning approaches for multi-omics data analysis: A review. *Biotechnology Advances*, page 107739, 2021. DOI: 10.1016/j.biotechadv.2021.107739 67

[278] Qingyi Yang, Asser Bassyouni, Christopher R. Butler, Xinjun Hou, Stephen Jenkinson, David A. Price, et al. Ligand biological activity predicted by cleaning positive and negative chemical correlations. *Proc. of the National Academy of Sciences*, 116(9):3373–3378, 2019. DOI: 10.1073/pnas.1810847116 67

[279] Steven Sandoval, Visar Berisha, Rene L. Utianski, Julie M. Liss, and Andreas Spanias. Automatic assessment of vowel space area. *The Journal of the Acoustical Society of America*, 134(5):EL477–EL483, 2013. DOI: 10.1121/1.4826150 68

[280] Visar Berisha, Julie Liss, Steven Sandoval, Rene Utianski, and Andreas Spanias. Modeling pathological speech perception from data with similarity labels. *IEEE International Conference on Acoustics, Speech and Signal Processing (ICASSP)*, pages 915–919, 2014. DOI: 10.1109/icassp.2014.6853730 68

[281] Dana Wagshal, Sethu Sankaranarayanan, Valerie Guss, Tracey Hall, Flora Berisha, Iryna Lobach, Anna Karydas, Lisa Voltarelli, Carole Scherling, Hilary Heuer, et al. Divergent CSF τ alterations in two common tauopathies: Alzheimer's disease and progressive supranuclear palsy. *Journal of Neurology, Neurosurgery and Psychiatry*, 86(3):244–250, 2015. DOI: 10.1136/jnnp-2014-308004 68

[282] Michael Esposito, Sunil Rao, Vivek Narayanaswamy, and Andreas Spanias. Covid-19 detection using audio spectral features and machine learning. *Asilomar Conference on Circuits, Systems and Computers*, Monterey, October 2021. 68

[283] Sunil Rao, Michael Esposito, Vivek Narayanaswamy, Jayaraman Thiagarajan, and Andreas Spanias. Deep learning with hyper-parameter tuning for covid-19 cough detection. *International Conference on Information, Intelligence, Systems and Applications (IISA)*, IEEE, July 2021. DOI: 10.1109/iisa52424.2021.9555564 68

[284] Mahesh Banavar, Henry Braun, Santoshi Tejasri Buddha, Venkatachalam Krishnan, Andreas Spanias, Shinichi Takada, Toru Takehara, Cihan Tepedelenlioglu, and Ted Yeider. Signal processing for solar array monitoring, fault detection, and optimization. *Synthesis Lectures on Power Electronics*, 7(1):1–95, 2012. DOI: 10.2200/s00425ed1v01y201206pel004 68

[285] Sameeksha Katoch, Gowtham Muniraju, Sunil Rao, Andreas Spanias, Pavan Turaga, Cihan Tepedelenlioglu, Mahesh Banavar, and Devarajan Srinivasan. Shading prediction, fault detection, and consensus estimation for solar array control. *IEEE Industrial Cyber-Physical Systems (ICPS)*, pages 217–222, 2018. DOI: 10.1109/icphys.2018.8387662 68

[286] Sunil Rao, David Ramirez, Henry Braun, Jongmin Lee, Cihan Tepedelenlioglu, Elias Kyriakides, Devarajan Srinivasan, Jeffrey Frye, Shinji Koizumi, Yoshitaka Morimoto, et al. An 18 KW solar array research facility for fault detection experiments. *18th Mediterranean Electrotechnical Conference (MELECON)*, pages 1–5, IEEE, 2016. DOI: 10.1109/melcon.2016.7495369 68

[287] Sunil Rao, Andreas Spanias, and Cihan Tepedelenlioglu. Solar array fault detection using neural networks. *IEEE International Conference on Industrial Cyber Physical Systems (ICPS)*, pages 196–200, 2019. DOI: 10.1109/icphys.2019.8780208 68

[288] Vivek Sivaraman Narayanaswamy, Raja Ayyanar, Andreas Spanias, Cihan Tepedelenlioglu, and Devarajan Srinivasan. Connection topology optimization in photovoltaic arrays

using neural networks. *IEEE International Conference on Industrial Cyber Physical Systems (ICPS)*, pages 167–172, 2019. DOI: 10.1109/icphys.2019.8780242 68

[289] Sunil Rao, Sameeksha Katoch, Vivek Narayanaswamy, Gowtham Muniraju, Cihan Tepedelenlioglu, Andreas Spanias, Pavan Turaga, Raja Ayyanar, and Devarajan Srinivasan. Machine learning for solar array monitoring, optimization, and control. *Synthesis Lectures on Power Electronics*, 7(1):1–91, Morgan & Claypool Publishers, 2020. DOI: 10.2200/s01027ed1v01y202006pel013 68

[290] Sunil Rao, Gowtham Muniraju, Cihan Tepedelenlioglu, Devarajan Srinivasan, Govindasamy Tamizhmani, and Andreas Spanias. Dropout and pruned neural networks for fault classification in photovoltaic arrays. *IEEE Access*, 2021. DOI: 10.1109/access.2021.3108684 68

[291] Glen Uehara, Sunil Rao, Mathew Dobson, Cihan Tepedelenlioglu, and Andreas Spanias. Quantum neural network parameter estimation for photovoltaic fault detection. *International Conference on Information, Intelligence, Systems and Applications (IISA)*, IEEE, July 2021. DOI: 10.1109/iisa52424.2021.9555558 68, 71, 72

[292] Kristen Jaskie, Joshua Martin, and Andreas Spanias. PV fault detection using positive unlabeled learning. *Applied Sciences*, 11(12):5599, 2021. DOI: 10.3390/app11125599 68

[293] Joshua Martin, Kristen Jaskie, Yiannis Tofis, and Andreas Spanias. PV array soiling detection using machine learning fault detection. *International Conference on Information, Intelligence, Systems and Applications (IISA)*, IEEE, July 2021. DOI: 10.1109/iisa52424.2021.9555535 68

[294] YuLong Zhang, ZiJie Dai, LongFei Zhang, ZhengYi Wang, Li Chen, and YuZhen Zhou. Application of artificial intelligence in military: From projects view. *6th International Conference on Big Data and Information Analytics (BigDIA)*, pages 113–116, 2020. DOI: 10.1109/bigdia51454.2020.00026 68

[295] Lakshya Tyagi, Vijay Kumar, and Subhecha Chakraborty. Explosion consequence analysis for military targets through support vector machines. *8th International Conference on Reliability, Infocom Technologies and Optimization (Trends and Future Directions) (ICRITO)*, pages 948–951, 2020. DOI: 10.1109/icrito48877.2020.9197866 68

[296] Yan Song, Yuemei Zhu, Guangliang Li, Chen Feng, Bo He, and Tianhong Yan. Side scan sonar segmentation using deep convolutional neural network. *OCEANS Anchorage*, pages 1–4, IEEE, 2017. 69

[297] Shanshan Song, Jun Liu, Jiani Guo, Jun Wang, Yanxin Xie, and Jun-Hong Cui. Neural-network-based AUV navigation for fast-changing environments. *IEEE Internet of Things Journal*, 7(10):9773–9783, 2020. DOI: 10.1109/jiot.2020.2988313 69

[298] Kazi Abu Taher, Billal Mohammed Yasin Jisan, and Md. Mahbubur Rahman. Network intrusion detection using supervised machine learning technique with feature selection. *International Conference on Robotics, Electrical and Signal Processing Techniques (ICREST)*, pages 643–646, 2019. DOI: 10.1109/icrest.2019.8644161 69

[299] Enrique V. Carrera, Fernando Lara, Marcelo Ortiz, Alexis Tinoco, and Rubén León. Target detection using radar processors based on machine learning. *IEEE ANDESCON*, pages 1–5, 2020. DOI: 10.1109/andescon50619.2020.9272173 69

[300] Xinyu Li, Yuan He, and Xiaojun Jing. A survey of deep learning-based human activity recognition in radar. *Remote Sensing*, 11(9):1068, 2019. DOI: 10.3390/rs11091068 69

[301] Jacquelyn Bulao. How much data is created every day in 2021? [you'll be shocked!], September 2021. 69

[302] TK Balaji, Chandra Sekhara Rao Annavarapu, and Annushree Bablani. Machine learning algorithms for social media analysis: A survey. *Computer Science Review*, 40:100395, 2021. DOI: 10.1016/j.cosrev.2021.100395 69

[303] Shiliang Sun, Chen Luo, and Junyu Chen. A review of natural language processing techniques for opinion mining systems. *Information Fusion*, 36:10–25, 2017. DOI: 10.1016/j.inffus.2016.10.004 70

[304] B. Senthil Arasu, B. Jonath Backia Seelan, and N. Thamaraiselvan. A machine learning-based approach to enhancing social media marketing. *Computers and Electrical Engineering*, 86:106723, 2020. DOI: 10.1016/j.compeleceng.2020.106723 70

[305] Anuja Arora, Shivam Bansal, Chandrashekhar Kandpal, Reema Aswani, and Yogesh Dwivedi. Measuring social media influencer index-insights from Facebook, Twitter and Instagram. *Journal of Retailing and Consumer Services*, 49:86–101, 2019. DOI: 10.1016/j.jretconser.2019.03.012 70

[306] Arkadiusz Paterek. *Predicting Movie Ratings and Recommender Systems: A 195-page Monograph on Machine Learning, Recommender Systems, and the Netflix Prize.* Arkadiusz Paterek, 2012. 70

[307] Salini Suresh, Niharika Sinha, Sabyasachi Prusty, et al. Latent approach in entertainment industry using machine learning. *International Research Journal on Advanced Science Hub*, 2:304–307, 2020. DOI: 10.47392/irjash.2020.106 70

[308] Michele Berno, Marco Canil, Nicola Chiarello, Luca Piazzon, Fabio Berti, Francesca Ferrari, Alessandro Zaupa, Nicola Ferro, Michele Rossi, and Gian Antonio Susto.

A machine learning-based approach for advanced monitoring of automated equipment for the entertainment industry. *IEEE International Workshop on Metrology for Industry 4.0 and IoT (MetroInd4. 0&IoT)*, pages 386–391, 2021. DOI: 10.1109/metroind4.0iot51437.2021.9488481 70

[309] Michael Dixon. How Netflix used big data and analytics to generate billions, September 2021. 71

[310] Chun-Ning Tsao, Jing-Kai Lou, and Homer H. Chen. Thumbnail image selection for VOD services. *IEEE Conference on Multimedia Information Processing and Retrieval (MIPR)*, pages 54–59, 2019. DOI: 10.1109/mipr.2019.00018 71

[311] Andrew Kusiak. Smart manufacturing. *International Journal of Production Research*, 56(1–2):508–517, 2018. DOI: 10.1080/00207543.2017.1351644 71

[312] László Monostori, András Márkus, Hendrik Van Brussel, and E Westkämpfer. Machine learning approaches to manufacturing. *CIRP Annals*, 45(2):675–712, 1996. DOI: 10.1016/s0007-8506(18)30216-6 71

[313] Duc T. Pham and Ashraf A. Afify. Machine-learning techniques and their applications in manufacturing. *Proc. of the Institution of Mechanical Engineers, Part B: Journal of Engineering Manufacture*, 219(5):395–412, 2005. DOI: 10.1243/095440505x32274 71

[314] James Clovis Kabugo, Sirkka-Liisa Jämsä-Jounela, Robert Schiemann, and Christian Binder. Industry 4.0 based process data analytics platform: A waste-to-energy plant case study. *International Journal of Electrical Power and Energy Systems*, 115:105508, 2020. DOI: 10.1016/j.ijepes.2019.105508 71

[315] Glen Uehara, Andreas Spanias and W. Clark. Quantum information processing algorithms with emphasis on machine learning. *International Conference on Information, Intelligence, Systems, and Applications (IISA)*, IEEE, July 2021. DOI: 10.1109/iisa52424.2021.9555570 71

[316] Seth Lloyd, Masoud Mohseni, and Patrick Rebentrost. Quantum algorithms for supervised and unsupervised machine learning. *ArXiv Preprint ArXiv:1307.0411*, 2013. 71

[317] Yuxiang Wang, Ruijin Wang, Dongfen Li, Daniel Adu-Gyamfi, Kaibin Tian, and Yixin Zhu. Improved handwritten digit recognition using quantum k-nearest neighbor algorithm. *International Journal of Theoretical Physics*, 58(7):2331–2340, 2019. DOI: 10.1007/s10773-019-04124-5 71

[318] Patrick Rebentrost, Masoud Mohseni, and Seth Lloyd. Quantum support vector machine for big data classification. *Physical Review Letters*, 113(13):130503, 2014. DOI: 10.1103/physrevlett.113.130503 71

[319] Marc Peter Deisenroth, A. Aldo Faisal, and Cheng Soon Ong. *Mathematics for Machine Learning.* Cambridge University Press, 2020. DOI: 10.1017/9781108679930 74

[320] Shai Shalev-Shwartz and Shai Ben-David. *Understanding Machine Learning: From Theory to Algorithms.* Cambridge University Press, 2014. DOI: 10.1017/cbo9781107298019 74

[321] Andreas C. Müller and Sarah Guido. *Introduction to Machine Learning with Python: A Guide for Data Scientists.* O'Reilly Media, Inc., 2016. 75

[322] Aurélien Géron. *Hands-on Machine Learning with Scikit-Learn, Keras, and TensorFlow: Concepts, Tools, and Techniques to Build Intelligent Systems.* O'Reilly Media, 2019. 75

[323] Eli Stevens, Luca Antiga, and Thomas Viehmann. *Deep Learning with PyTorch.* Manning Publications Company, 2020. 75

[324] Francois Chollet et al. *Deep Learning with Python*, vol. 361. Manning New York, 2018. 75

[325] Tom B. Brown, Benjamin Mann, Nick Ryder, Melanie Subbiah, Jared Kaplan, Prafulla Dhariwal, Arvind Neelakantan, Pranav Shyam, Girish Sastry, Amanda Askell, et al. Language models are few-shot learners. *ArXiv Preprint ArXiv:2005.14165*, 2020.

[326] Joan Bruna, Wojciech Zaremba, Arthur Szlam, and Yann LeCun. Spectral networks and locally connected networks on graphs. *Proc. of the 2nd International Conference on Learning Representations, ICLR'14*, pages 1–14, Banff, Canada, 2014. 38

[327] David K. Duvenaud, Dougal Maclaurin, Jorge Iparraguirre, Rafael Bombarell, Timothy Hirzel, Alán Aspuru-Guzik, and Ryan P. Adams. Convolutional networks on graphs for learning molecular fingerprints. *Advances in Neural Information Processing Systems*, pages 2224–2232, 2015. 38

[328] Will Hamilton, Zhitao Ying, and Jure Leskovec. Inductive representation learning on large graphs. *Advances in Neural Information Processing Systems*, pages 1024–1034, 2017.

[329] Victoria J. Hodge, Simon O'Keefe, Michael Weeks, and Anthony Moulds. Wireless sensor networks for condition monitoring in the railway industry: A survey. *IEEE Transactions on Intelligent Transportation Systems*, 16(3):1088–1106, 2014. DOI: 10.1109/tits.2014.2366512

[330] Mathias Niepert, Mohamed Ahmed, and Konstantin Kutzkov. Learning convolutional neural networks for graphs. *International Conference on Machine Learning*, pages 2014–2023, 2016. 38

[331] Petar Velickovic, Guillem Cucurull, Arantxa Casanova, Adriana Romero, Pietro Lio, and Yoshua Bengio. Graph attention networks. *ArXiv Preprint ArXiv:1710.10903*, 2017. 38

[332] Anant S. Vemuri. Survey of computer vision and machine learning in gastrointestinal endoscopy. *ArXiv Preprint ArXiv:1904.13307*, 2019.

[333] Thorsten Wuest, Daniel Weimer, Christopher Irgens, and Klaus-Dieter Thoben. Machine learning in manufacturing: Advantages, challenges, and applications. *Production and Manufacturing Research*, 4(1):23–45, 2016. DOI: 10.1080/21693277.2016.1192517

[334] Zhilu Zhang and Mert R. Sabuncu. Generalized cross entropy loss for training deep neural networks with noisy labels. *ArXiv Preprint ArXiv:1805.07836*, 2018. 10

Authors' Biographies

UDAY SHANKAR SHANTHAMALLU

Uday Shankar Shanthamallu received his Ph.D. degree in 2021 from the school of Electrical, Computer, and Energy Engineering at Arizona State University. He received his Master's degree in electrical engineering from Arizona State University in 2018 and a Bachelor's degree in electronics and communication engineering from the National Institute of Engineering, India, in 2011. His research interests include representation learning for graphs using machine learning and deep learning techniques. He also has experience on sensor data analytics for anomaly detection. His internship with NXP Semiconductors (2016) focused on algorithm development for sensor data analytics. He also interned with Lawrence Livermore National Laboratory during the summer of 2019 and 2020 where he built predictive models for human brain connectomes.

ANDREAS SPANIAS

Andreas Spanias is Professor in the School of Electrical, Computer, and Energy Engineering at Arizona State University. He is also the director of the Sensor Signal and Information Processing (SenSIP) center and the founder of the SenSIP industry consortium (also an NSF I/UCRC site). His research interests are in the areas of adaptive signal processing, speech processing, machine learning and sensor systems. He and his student team developed the computer simulation software Java-DSP and its award-winning iPhone/iPad and Android versions. He is author of two textbooks: *Audio Processing and Coding* by Wiley and *DSP: An Interactive Approach* (2nd ed.). He contributed to more than 350 papers, 11 monographs, 11 full patents, 10 provisional patents, and 12 patent pre-disclosures. He served as Associate Editor of the *IEEE Transactions on Signal Processing* and as General Co-chair of *IEEE ICASSP*-99. He also served as the IEEE Signal Processing Vice-President for Conferences. Andreas Spanias is co-recipient of the 2002 IEEE Donald G. Fink paper prize award and was elected Fellow of the IEEE in 2003. He served as Distinguished Lecturer for the IEEE Signal Processing Society in 2004. He is a series editor for the Morgan & Claypool lecture series on algorithms and software. He received the 2018 IEEE Phoenix Chapter award with citation: "For significant innovations and patents in signal processing for sensor systems." He also received the 2018 IEEE Region 6 Outstanding Educator Award (across 12 states) with citation: "For outstanding research and education contributions in signal processing." He was elected recently to Senior Member of the National Academy of Inventors (NAI).

Printed in the United States
by Baker & Taylor Publisher Services

Printed in the United States
by Baker & Taylor Publisher Services